Practical Atom Model

David Michalets

I0491636

Self-published on **April 11, 2021**

Table of Contents

Introduction

This book proposes a practical atomic model, having less uncertainty with predictable behaviors. Protons and electrons are real particles having certain defined and predictable behaviors. An atom's rings of electrons are certainly predictable or else the frequent practice of predicting the results of chemical reactions with certainty would be impossible.

An atom is capable of adapting to changes. These include a) the nucleus which can increase through fusion and transmutation, and b) the electron cloud which must increase its quantity when the nucleus adds protons.

Instability in the nucleus forces an action observed as radioactive decay.

These behaviors involve mechanisms driven by the electrostatic forces between the positive and negative charged particles within the atom. Whenever equilibrium is disturbed, a spontaneous reaction occurs to restore it.

The consistent organization of an atom's electrons is important. One emphasis of this book is explaining the structure within the electron cloud around the nucleus. The current concept of quantum orbitals for the pdf shells, claimed to be driven by probabilities and a supposed wave behavior, is inconsistent with the predictability and certainty present for chemists to manage chemical reactions. The electrons in the cloud cannot be moving subject to probabilities. All electrons are in a predictable pattern in specific rings around the nucleus.

The Bohr atomic model was loosely based on the solar system but having electrons in circular orbits (not elliptical), but later replaced with the current theory of "spdf" orbitals.

This practical atomic model seeks to address the major issues of an atomic model, including behaviors in the electron configuration, and behaviors in the nucleus, like its decay and its mass defect.

An attempt is made to distinguish the relative importance of some observations. For example, quarks add nothing to our understanding of atomic behaviors.

The Bohr model having circular orbits must be reconsidered. Each atom's electrons appear to be in predictable rings. When electrons are predictable, they cannot have a probability driven behavior, described as a wave, at a larger scale.

The valence behavior of the elements will be addressed.

The author's previous book, Practical Particle Physics (PPP), established the Standard Model is an incorrect model having several inadequate explanations. The main problem solved in that book is explaining the atomic mass defect. The Standard Model does not understand the fundamental particles in a nucleus correctly, causing the model's inability to explain an atom's measured mass value. A change is needed when describing an atom's nucleus This book is the logical sequel, by describing updates to the defective Standard Model where needed. Some were identified in that previous book, and more conclusions are presented in this one.

Each element has 2 important behaviors for its atoms, 1) actions during radioactive decay among its unstable isotopes, and 2) maintaining a collection of electrons in several shells, which enable one atom maintaining chemical bonds with other atoms.

Sometimes during several events, the nucleus interacts with its electrons. PPP is not required to read this book because some critical content from PPP is adapted here.

This is a brief summary of the 12 sections:

1)   Background describes several terms and behaviors. These topics include Planck's constant, the Rydberg constant, and definitions regarding electron shell configurations.

2)   Fundamentals describes several fundamental terms and assumptions, which support the conclusions being derived later in the book.

Much of this section is the foundation of particle physics, with its important facets presented.

3)   Data Sets describes several compilations by the author of several measurements among the 118 elements. The author's conclusions require evidence.

4)   Build Nucleus describes the process of building a nucleus using protons and electrons.

5)   Gather Electrons describes the predictable scheme for each element's electrons, including the valence electrons; a change in the naming convention for the rings is considered.

6)   Atomic Bonds describes the atomic behaviors in the electron shells when atoms interact by forming bonds using shared electrons to create compounds or molecules. These bonds also include a crystal which is a lattice structure of atomic nuclei, without distinct molecules.

7)   Radioactive Decay describes the respective behaviors during a decay sequence.

8)   Periodic Table describes several behaviors which vary among the elements in the periodic table. There are 118 elements.

For each:

a) The electron configuration is provided which also defines an atom's valence.

b) The element's stable isotopes or those radioactive isotopes having a notable half-life are provided.

c) Any anomalies in the electron configuration are noted.

9)   Atomic Model describes changes to the current atomic model,

10)   Light describes the interactions of light with atoms and matter.

11)   Final Conclusion summarizes the book's conclusions.

12)   All external references in the book have links available as directed here.

1 Background

There are several topics which are mentioned later, so they are described here, in the beginning.

1.1 Author's Preceding Books

This is the author's sixth book.

It builds on some of the work and conclusions already published. None of those earlier efforts are required to read this book. However, it can be helpful knowing the fifth book established part of the foundation for this sixth book. Most of its details are not repeated here but they were the basis for further analysis of data compiled after completing the earlier works.

The previous 5 books are summarized next.

1.1.1 Observing Our Universe

This, OOU, is the author's first book. It addressed many issues in cosmology, especially the mistakes made with a spectrum analysis of distant objects, such as galaxies and quasars. There are actually 4 different red shift mechanisms. Treating them as all the same results in many mistakes. There is no universe expansion and there was no big bang. There are two other problems when ignoring the observer's context.

One is the special observer in relativity; no celestial objects in cosmology have such a special observer commanding their motion in their reference frame.

The second is the detection of non-existent gravitational waves. LIGO made a mistake when their detectors on Earth's surface reacted to tidal stress by the Moon or Sun, but LIGO falsely claimed a distant astrophysical source.

The author gave LIGO predictions in November, 2019, which were confirmed by LIGO detections as expected. The exercise confirmed LIGO's mistake. LIGO never detected anything being claimed. They never provided evidence for their claims. LIGO is an abomination.

1.1.2 Cosmology Transition

This, CT, is the author's second book. It addressed many issues in cosmology, which were identified in the first book.

As part of the transition, a better scheme for archiving astronomical data is proposed.

A new solar model, proposed by Pierre-Marie Robitaille, is described.

This is a paradigm shift with stars having a hot solid core, cooling by convection out to the photosphere which cools by radiating a thermal spectrum. This model matches all solar observations, including the liquid surface of the photosphere. There is no unstable internal fusion mechanism.

The SAFIRE project duplicated the conditions of the photosphere in a controlled experiment and observed element transmutation on the globe's surface. Elements are created on the stellar surface, not in the core.

A new spiral galaxy model, proposed by Donald Scott, is also described. This model explains the disk's rotation curve by the galaxy's magnetic field, eliminating the mistake of dark matter.

The author's new quasar model is described. Halton Arp did not understand the mechanism driving a measured red shift.

The author also describes the basics of the entity the author calls a sphere of stars. These include a globular cluster and elliptical galaxy. Several conjectures are offered but more public data are required for better explanations of these distant objects.

1.1.3 Cosmology Connections

This, CC, is the author's third book. It addressed several issues in cosmology, especially when missing electrical connections among celestial bodies. There is an important distinction between thermal radiation and synchrotron radiation.

1.1.4 Redefining Gravity

This, RG, is the author's fourth book. It describes the author's mechanism for Newton's force of gravity. It describes the many confirmations of the force, including Kepler's laws of planetary motion. It also explains the issues of space-time. Gravity must be redefined to be a real, fundamental force, not as curvature.

1.1.5 Practical Particle Physics

This, PPP, is the author's fifth book. It included an extensive research into the atomic mass defect behavior (that analysis is improved in this sixth book). Its conclusions included changes to the atomic model, including the proton. All of the atomic behaviors with light were described including: particle pair production, Compton scattering, photoelectric effect, and Doppler Effect. All can be explained with no photon because all are a wave length driven behavior. Electron shell configurations were not explained. That is a significant part of this sixth book.

Sometimes, PPPB is used for Practical Particle Physics Book, as a compact reference when its conclusions are described here, in the sixth book.

PPPB is not required reading for this sixth book, though its analysis of quarks, gravitons, photons and several atomic behaviors could be helpful background.
Quarks are debris found only in particle accelerators. Muons can be found after the interstellar particle accelerator known as cosmic rays. They, like quarks, must be given the classification of debris and are not relevant to an atomic model.

Visible Light is part of the wave length continuum making up part of the electromagnetic radiation spectrum, which is possible from the propagation of synchronized, perpendicular electric and magnetic fields whose oscillation can be measured as either frequency or wave length.

There is no photon, though this quasi-particle is assumed a valid particle in current particle physics.

1.2 Planck's Constant

Planck's constant is very important in particle physics.

Lori Gardi recently concluded Planck's equation has a mistake in its units where h x f cannot = energy without a change.
Links to the academic paper and video are in References.

She had other important conclusions. The title of her paper is:

Planck's Constant and the Nature of Light

The YouTube video of that title by Lori Gardi and the paper are recommended for 5 reasons:

1) The well accepted Planck's equation has a bug. She has a thorough explanation, which is worthwhile to hear.

A simple observation worth noting here is:

Planck's equation has a mistake in its units. The constant h has seconds but the equation has no variable for time. The equation needs a fix.

E = hf should become either:
a) ΔE = hf   or

b) E = htf

where t is the time for the measurement with this frequency.

2) The energy in light is in the intensity of a particular wave length, not only in the frequency, as is currently implied by the mistaken formula.

3) This video is another useful explanation of why there is no photon.

Light is a wave, not a particle. The synchronized electric and magnetic fields are oscillating with a consistent wave length during its propagation after initiation. Light never has a particle behavior. In an updated, practical atomic model, there are no fictitious quasi-particles like a photon.

Particles require a mass to be detectable and measurable.

4) Planck's constant defines the minimum measurable wave length of light.

In some cases, the usage of Planck's constant must change because its units failed to address the missing time variable in Planck's equation. One usage is the uncertainty principle.

5) The uncertainty principle in quantum mechanics can have the uncertain limits defined so now they are not truly uncertain.

She proposes a new Planck's constant, h_q having no seconds in its units but with the same value as h. The q subscript identifies this h_q as the "quantum of energy." The uncertainty principle now has less uncertainty because it has a defined limit:

The minimum detectable change in energy must be greater than or equal to h_q.

Copying much of her content here is not appropriate.

I have nothing to contribute to her excellent work. Excerpts can remove important context, in a case like this.

Observation:

From the video, this equation should be true:

$$\Delta E = hc/\lambda$$

Because a wave length is often used in this book, this particular formula is important.
The term "quantum of energy" for one photon refers to this equation. There is no photon, but a wave length can carry energy as calculated here.

1.2.1 The Reduced Planck's Constant

Planck's constant has a "reduced value" and its symbol is called h-bar.

Excerpt from Wikipedia:

The Planck constant has dimensions of physical action; i.e., energy multiplied by time, or momentum multiplied by distance, or angular momentum. In SI units, the Planck constant is expressed in joule-seconds (J*s or $N \cdot ms$ or $kg \cdot m^2/s$). Implicit in the dimensions of the Planck constant is the fact that the SI unit of frequency, the hertz, represents one complete cycle, 360 degrees or 2π radians, per second. An angular frequency in radians per second is often more natural in mathematics and physics and many formulas use a reduced Planck constant (pronounced h-bar) -- thus apparently just J*s/cycle and J*s/radian units.

$$h = 6.626\ 07015 \times 10^{-34} \ J \cdot s$$

$$h\text{-bar} = h\ /\ 2\pi = 1.05405710817...x\ 1034 \ x10^{-34} \ J \cdot s =$$
$$6.582\ 119\ 569... \ x10^{-16} \ eV \cdot s$$

(Excerpt end)

Observation:

The h-bar value has seconds in its units, like h.

1.3 Rydberg Constant

The Rydberg constant is important in particle physics.

Excerpt from Wikipedia:

In spectroscopy, the Rydberg constant is a physical constant relating to the electromagnetic spectra of an atom. The constant first arose as an empirical fitting parameter in the Rydberg formula for the hydrogen spectral series, but Niels Bohr later showed that its value could be calculated from more fundamental constants via his Bohr model. As of 2018, [this constant] and electron spin g-factor are the most accurately measured physical constants.

The constant is expressed for either hydrogen as R_H or at the limit of infinite nuclear mass as R_∞. In either case, the constant is used to express the limiting value of the highest wavenumber (inverse wavelength) of any photon that can be emitted from an atom, or, alternatively, the wavenumber of the lowest-energy photon capable of ionizing an atom from its ground state. The hydrogen spectral series can be expressed simply in terms of the Rydberg constant for hydrogen RH and the Rydberg formula.

In atomic physics, Rydberg unit of energy, symbol Ry, corresponds to the energy of the photon whose wavenumber is the Rydberg constant, i.e. the ionization energy of the hydrogen atom in a simplified Bohr model.

(Excerpt end)

Observation:

The formula for the Rydberg constant uses only me not m_p. In PPPB, m_p has a recommended reduction to make sure: $m_p + m_e$ = mass of ^{1}H.

There should be no justification for that simple equation of masses to be wrong when both ^{1}H and m_e are correct.

Apparently, that recommended change in m_p cannot affect the Rydberg constant.

However, the formula for the Rydberg constant for hydrogen uses both m_e and m_p.

This means R_H could be reduced, but the change in m_p had no change in value with the 10 digit precision of my MS Excel.

Maybe this consistency should not be a surprise. The new m_p mass was based on the well-established electron-proton mass ratio.

1.4 Electron Shell

This function is critical in this book.

Excerpt from Wikipedia:

In chemistry and atomic physics, an electron shell may be thought of as an orbit followed by electrons around an atom's nucleus. The closest shell to the nucleus is called the "1 shell" (also called the "K shell"), followed by the "2 shell" (or "L shell"), then the "3 shell" (or "M shell"), and so on farther and farther from the nucleus. The shells correspond to the principal quantum numbers ($n = 1, 2, 3, 4 ...$) or are labeled alphabetically with the letters used in X-ray notation (K, L, M, ...).

Each shell can contain only a fixed number of electrons: The first shell can hold up to two electrons, the second shell can hold up to eight ($2 + 6$) electrons, the third shell can hold up to 18 ($2 + 6 + 10$) and so on. The general formula is that the nth shell can in principle hold up to $2(n^2)$ electrons. For an explanation of why electrons exist in these shells see electron configuration.

Each shell consists of one or more subshells, and each subshell consists of one or more atomic orbitals.

(Excerpt end)

Observation:

Electron Configuration is mentioned and is next.

1.5 Electron Configuration.

This function is critical in this book.

Excerpt from Wikipedia:

In atomic physics and quantum chemistry, the electron configuration is the distribution of electrons of an atom or molecule (or other physical structure) in atomic or molecular orbitals. For example, the electron configuration of the neon atom is 1s2 2s2 2p6.

Electronic configurations describe each electron as moving independently in an orbital, in an average field created by all other orbitals. Mathematically, configurations are described by Slater determinants or configuration state functions.

According to the laws of quantum mechanics, for systems with only one electron, a level of energy is associated with each electron configuration and in certain conditions, electrons are able to move from one configuration to another by the emission or absorption of a quantum of energy, in the form of a photon.

(Excerpt end)

Observation:

As noted in an earlier section, "quantum of energy" is the energy in one wave length of light.

1.6 Energy Level

An atom holds energy in its electrons.

Excerpt from Wikipedia:

A quantum mechanical system or particle that is bound—that is, confined spatially—can only take on certain discrete values of energy, called energy levels. This contrasts with classical particles, which can have any amount of energy. The term is commonly used for the energy levels of the electrons in atoms, ions, or molecules, which are bound by the electric field of the nucleus, but can also refer to energy levels of nuclei or vibrational or rotational energy levels in molecules. The energy spectrum of a system with such discrete energy levels is said to be quantized. (And the energy levels don't have to be equal)
In chemistry and atomic physics, an electron shell, or principal energy level, may be thought of as the orbit of one or more electrons around an atom's nucleus. The closest shell to the nucleus is called the "1 shell" (also called "K shell"), followed by the "2 shell" (or "L shell"), then the "3 shell" (or "M shell"), and so on farther and farther from the nucleus. The shells correspond with the principal quantum numbers (n = 1, 2, 3, 4 ...) or are labeled alphabetically with letters used in the X-ray notation (K, L, M, N...).

Each shell can contain only a fixed number of electrons: The first shell can hold up to two electrons, the second shell can hold up to eight (2 + 6) electrons, the third shell can hold up to 18 (2 + 6 + 10) and so on. The general formula is that the nth shell can in principle hold up to $2(n^2)$ electrons. Since electrons are electrically attracted to the nucleus, an atom's electrons will generally occupy outer shells only if the more inner shells have already been completely filled by other electrons. However, this is not a strict requirement: atoms may have two or even three incomplete outer shells. For an explanation of why electrons exist in these shells see electron configuration.

If the potential energy is set to zero at infinite distance from the atomic nucleus or molecule, the usual convention, then bound electron states have negative potential energy. If an atom, ion, or molecule is at the lowest possible energy level, it and its electrons are said to be in the ground state. If it is at a higher energy level, it is said to be excited, or any electrons that have higher energy than the ground state are excited. If more than one quantum mechanical state is at the same energy, the energy levels are "degenerate". They are then called degenerate energy levels.

(Excerpt end)

1.7 Momentum

Momentum is defined before angular momentum, which follows.

Excerpt from Wikipedia:

In Newtonian mechanics, linear momentum, translational momentum, or simply momentum (pl. momenta) is the product of the mass and velocity of an object. It is a vector quantity, possessing a magnitude and a direction. If m is an object's mass and v is its velocity (also a vector quantity), then the object's momentum is:
$p = mv$
In SI units, momentum is measured in kilogram meters per second (kg · m/s).

(Excerpt end)

1.8 Angular Momentum

This function is relevant to the electron cloud.

Excerpt from Wikipedia:

In physics, angular momentum (rarely, moment of momentum or rotational momentum) is the rotational equivalent of linear momentum. It is an important quantity in physics because it is a conserved quantity—the total angular momentum of a closed system remains constant.

In three dimensions, the angular momentum for a point particle is a pseudovector $r \times p$, the cross product of the particle's position vector r (relative to some origin) and its momentum vector; the latter is $p = mv$ in Newtonian mechanics. This definition can be applied to each point in continua like solids or fluids, or physical fields. Unlike momentum, angular momentum does depend on where the origin is chosen, since the particle's position is measured from it.

Just like for angular velocity, there are two special types of angular momentum: the spin angular momentum and orbital angular momentum. The spin angular momentum of an object is defined as the angular momentum about its centre of mass coordinate. The orbital angular momentum of an object about a chosen origin is defined as the angular momentum of the centre of mass about the origin. The total angular momentum of an object is the sum of the spin and orbital angular momenta. The orbital angular momentum vector of a point particle is always parallel and directly proportional to the orbital angular velocity vector ω of the particle, where the constant of proportionality depends on both the mass of the particle and its distance from origin. The spin angular momentum vector of a rigid body is proportional but not always parallel to the spin angular velocity vector Ω, making the constant of proportionality a second-rank tensor rather than a scalar.

Angular momentum is an extensive quantity; i.e. the total angular momentum of any composite system is the sum of the angular momenta of its constituent parts. For a continuous rigid body, the total angular momentum is the volume integral of angular momentum density (i.e. angular momentum per unit volume in the limit as volume shrinks to zero) over the entire body.

Torque can be defined as the rate of change of angular momentum, analogous to force. The net external torque on any system is always equal to the total torque on the system; in other words, the sum of all internal torques of any system is always 0 (this is the rotational analogue of Newton's Third Law). Therefore, for a closed system (where there is no net external torque), the total torque on the system must be 0, which means that the total angular momentum of the system is constant. The conservation of angular momentum helps explain many observed phenomena, for example the increase in rotational speed of a spinning figure skater as the skater's arms are contracted, the high rotational rates of neutron stars, the Coriolis effect, and the precession of gyroscopes. In general, conservation does limit the possible motion of a system, but does not uniquely determine what the exact motion is.

In quantum mechanics, angular momentum (like other quantities) is expressed as an operator, and its one-dimensional projections have quantized eigenvalues. Angular momentum is subject to the Heisenberg uncertainty principle, implying that at any time, only one projection (also called "component") can be measured with definite precision; the other two then remain uncertain.

Because of this, the notion of a quantum particle literally "spinning" about an axis does not exist. Quantum particles do possess a type of non-orbital angular momentum called "spin", but this angular momentum does not correspond to actual physical spinning motion.

(Excerpt end)

1.9 Nucleus

An atom's nucleus contains protons and neutrons.
In the author's PPPB, new behaviors for protons and
neutrons are described. This book assumes those
conclusions are valid, and that analysis is not repeated
here.

In summary, protons compressed into contact cause the
repulsive electrostatic force to reverse becoming attractive.
The reversal returns to repulsive if equilibrium is disturbed.
Having negative electrons attached to protons in the
nucleus defines the strong force where the proximity of
opposing charges maintains the mutual attraction.

The topology of the particles in the nucleus affects its
stability. There is a data set for this research of
combinations for a stable nucleus.

1.10 Electronegativity

This function is relevant in this book.

Excerpt from Wikipedia:

Electronegativity, symbol χ, measures the tendency of
an atom to attract a shared pair of electrons (or electron
density). An atom's electronegativity is affected by both
its atomic number and the distance at which its valence
electrons reside from the charged nucleus. The higher the
associated electronegativity, the more an atom or a
substituent group attracts electrons.
On the most basic level, electronegativity is determined by
factors like the nuclear charge (the more protons an atom
has, the more "pull" it will have on electrons) and the

number and location of other electrons in the atomic shells (the more electrons an atom has, the farther from the nucleus the valence electrons will be, and as a result, the less positive charge they will experience—both because of their increased distance from the nucleus and because the other electrons in the lower energy core orbitals will act to shield the valence electrons from the positively charged nucleus).

The opposite of electronegativity is electropositivity: a measure of an element's ability to donate electrons.

The term "electronegativity" was introduced by Jöns Jacob Berzelius in 1811, though the concept was known before that and was studied by many chemists including Avogadro. In spite of its long history, an accurate scale of electronegativity was not developed until 1932, when Linus Pauling proposed an electronegativity scale which depends on bond energies, as a development of valence bond theory. It has been shown to correlate with a number of other chemical properties. Electronegativity cannot be directly measured and must be calculated from other atomic or molecular properties. Several methods of calculation have been proposed, and although there may be small differences in the numerical values of the electronegativity, all methods show the same periodic trends between elements.

The most commonly used method of calculation is that originally proposed by Linus Pauling. This gives a dimensionless quantity, commonly referred to as the Pauling scale (χr), on a relative scale running from 0.79 to 3.98 (hydrogen = 2.20). When other methods of calculation are used, it is conventional (although not obligatory) to quote the results on a scale that covers the same range of numerical values: this is known as an electronegativity in Pauling units.

As it is usually calculated, electronegativity is not a property of an atom alone, but rather a property of an atom in a molecule. Properties of a free atom include ionization energy and electron affinity. It is to be expected that the electronegativity of an element will vary with its chemical environment, but it is usually considered to be a transferable property, that is to say that similar values will be valid in a variety of situations.

Caesium is the least electronegative element (0.79); fluorine is the most (3.98).

(Excerpt end)

Observation:

This book cannot be a chemistry text book. One emphasis is on the electron confirmation and its rings. Certain topics are covered in detail while others covered in detail by chemistry are only noted here when appropriate. A practical atomic model must recognize the bigger context from chemistry, to avoid too much attention on subatomic particle behaviors, especially the quasi-particles.

1.11 Valence Electron

This concept is important to atomic bonds.

Excerpt from Wikipedia:

Valence electron is an outer shell electron that is associated with an atom, and that can participate in the formation of a chemical bond if the outer shell is not closed; in a single covalent bond, both atoms in the bond contribute one valence electron in order to form a shared pair.
The presence of valence electrons can determine the element's chemical properties, such as its valence— whether it may bond with other elements and, if so, how readily and with how many. In this way, a given element's reactivity is highly dependent upon its electronic configuration.

For a main group element, a valence electron can exist only in the outermost electron shell; for a transition metal, a valence electron can also be in an inner shell.
An atom with a closed shell of valence electrons (corresponding to an electron configuration s2p6 for main group elements or d10s2p6 for transition metals) tends to be chemically inert. Atoms with one or two valence electrons more than a closed shell are highly reactive due to the relatively low energy to remove the extra valence electrons to form a positive ion. An atom with one or two electrons less than a closed shell is reactive due to its tendency either to gain the missing valence electrons and form a negative ion, or else to share valence electrons and form a covalent bond.

Similar to a core electron, a valence electron has the ability to absorb or release energy in the form of a photon. An energy gain can trigger the electron to move (jump) to an outer shell; this is known as atomic excitation. Or the electron can even break free from its associated atom's shell; this is ionization to form a positive ion. When an electron loses energy (thereby causing a photon to be emitted), then it can move to an inner shell which is not fully occupied.

(Excerpt end)

Observation:

The number of valence electrons is not always determined by the outer shell. This number will be identified for each element.

1.12 Ionization Energy

This is related to valence and is covered in more detail in the section 9 Atomic Model.

1.13 Atomic orbital

This term is used frequently in the Standard Model.

Excerpt from Wikipedia:

In atomic theory and quantum mechanics, an atomic orbital is a mathematical function describing the location and wave-like behavior of an electron in an atom. This function can be used to calculate the probability of finding any electron of an atom in any specific region around the atom's nucleus. The term atomic orbital may also refer to the physical region or space where the electron can be calculated to be present, as predicted by the particular mathematical form of the orbital.

(Excerpt end)

Orbitals are explained in detail in section 5 Gather Electrons.

1.14 Atomic Theory

A little background on the current atomic model is useful.

Excerpt from Wikipedia:

In the early 1800s, the scientist John Dalton noticed that chemical substances seemed to combine and break down into other substances by weight in proportions that suggested that each chemical element is ultimately made up of tiny indivisible particles of consistent weight.

Shortly after 1850, certain physicists developed the kinetic theory of gases and of heat, which mathematically modeled the behavior of gases by assuming that they were made of particles.

In the early 20th century, Albert Einstein and Jean Perrin proved that Brownian motion (the erratic motion of pollen grains in water) is caused by the action of water molecules; this third line of evidence silenced remaining doubts among scientists as to whether atoms and molecules were real. Throughout the nineteenth century, some scientists had cautioned that the evidence for atoms was indirect, and therefore atoms might not actually be real, but only seem to be real.

By the early 20th century, scientists had developed fairly detailed and precise models for the structure of matter, which led to more rigorously-defined classifications for the tiny invisible particles that make up ordinary matter. An atom is now defined as the basic particle that composes a chemical element. Around the turn of the 20th century, physicists discovered that the particles that chemists called "atoms" are in fact agglomerations of even smaller particles (subatomic particles), but scientists kept the name out of convention. The term elementary particle is now used to refer to particles that are actually indivisible.

(Excerpt end)

Wikipedia has an image for this testing of competing atomic models:

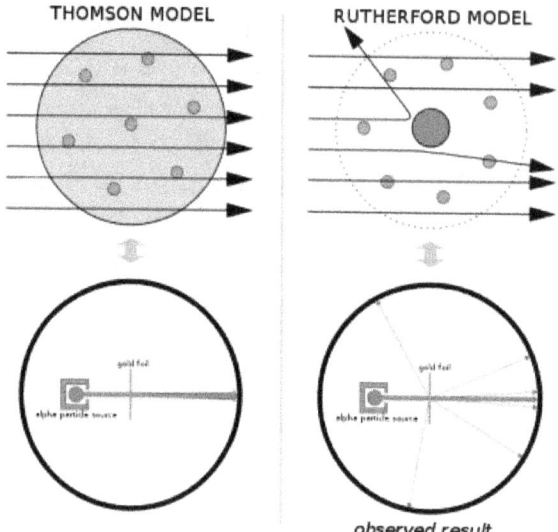

Its caption:

The Geiger–Marsden experiment
Left: Expected results: alpha particles passing through the plum pudding model of the atom with negligible deflection.
Right: Observed results: a small portion of the particles were deflected by the concentrated positive charge of the nucleus.

Observation:

The alpha particles were affected by only the nucleus, not the electrons around it. Gold foil was used. Gold has 79 electrons in an atom. There was nothing to detect whether any electrons were affected by the intruding alpha particle.

Excerpt continues with the new model.

In 1924, Louis de Broglie proposed that all moving particles—particularly subatomic particles such as electrons—exhibit a degree of wave-like behavior. Erwin Schrödinger, fascinated by this idea, explored whether or not the movement of an electron in an atom could be better explained as a wave rather than as a particle. Schrödinger's equation, published in 1926, describes an electron as a wave function instead of as a point particle. This approach elegantly predicted many of the spectral phenomena that Bohr's model failed to explain. Although this concept was mathematically convenient, it was difficult to visualize, and faced opposition. One of its critics, Max Born, proposed instead that Schrödinger's wave function described not the electron but rather all its possible states, and thus could be used to calculate the probability of finding an electron at any given location around the nucleus. This reconciled the two opposing theories of particle versus wave electrons and the idea of wave–particle duality was introduced. This theory stated that the electron may exhibit the properties of both a wave and a particle. For example, it can be refracted like a wave, and has mass like a particle.

A consequence of describing electrons as waveforms is that it is mathematically impossible to simultaneously derive the position and momentum of an electron. This became known as the Heisenberg uncertainty principle after the theoretical physicist Werner Heisenberg, who first described it and published it in 1927. This invalidated Bohr's model, with its neat, clearly defined circular orbits. The modern model of the atom describes the positions of electrons in an atom in terms of probabilities. An electron can potentially be found at any distance from the nucleus, but, depending on its energy level, exists more frequently in certain regions around the nucleus than others; this pattern is referred to as its atomic orbital.

The orbitals come in a variety of shapes-
sphere, dumbbell, torus, etc.-with the nucleus in the middle.

(Excerpt end)

Wikipedia provides this image with examples:

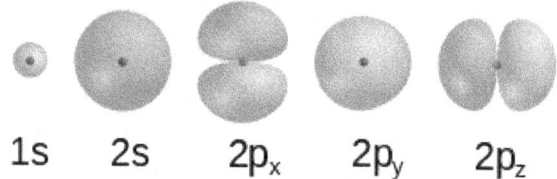

1s   2s   2p$_x$ 2p$_y$ 2p$_z$

Caption:
The five filled atomic orbitals of a neon atom separated and
arranged in order of increasing energy from left to right, with
the last three orbitals being equal in energy. Each orbital
holds up to two electrons, which most probably exist in the
zones represented by the colored bubbles. Each electron is
equally present in both orbital zones, shown here by color
only to highlight the different wave phase.

Observation:

If electrons were truly in such strange orbital paths, a
predictable covalent bond between atoms would be difficult.
The two nuclei must share an electron, or multiple, so the
valence shell must be at the edge of the electron cloud.
These internal paths would prevent them making a bond,
between the combinations of orbital paths in the separate
atoms.

Wikipedia (Atomic orbital) has this example for sodium
atom. Though claiming to be sodium, the image actually
represents neon having 1s2, 2s2, 2p6.

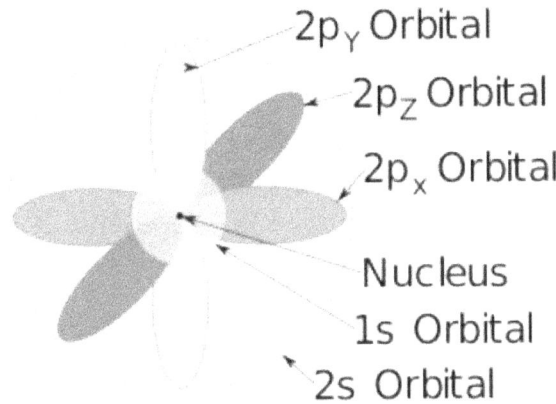

2p$_Y$ Orbital

2p$_Z$ Orbital

2p$_X$ Orbital

Nucleus

1s Orbital

2s Orbital

Caption:

The 1s, 2s, & 2p orbitals of a sodium atom.

Observation:

Bohr's model supposedly had unacceptable circular orbits requiring unacceptable velocities. The solution was proposing electrons moving among undefined locations, uncertain by Heisenberg, so any velocity is quite impossible to calculate. This solution offers an unverifiable explanation for a claimed slower velocity.

Perhaps the electrons are not truly moving in these strange orbits at whatever velocity is required for that proposed non-circular path.

Whether electrons are moving in a predictable orbit within the cloud is the fundamental issue in an atomic model. The electron cloud is a major topic of this book.

The image for the sodium atom is notable because only the 1s and 2s orbitals are circular. The electrons in the 2p orbital are not circular.

The 6 electrons in the 2p orbital are at the 2s orbital for only a brief instant according to this representation, while chemical reactions require them to be ALWAYS at the 2s circular orbital.

The image for sodium is notable for a second reason: the 3s shell of sodium is missing.

For its atomic orbital topic, Wikipedia presents and describes the neon shell configuration while calling it sodium.
Presenting the shell configuration for sodium in a manner like neon could present a problem with scaling.

The covalent radius of neon is 58 pm.
The covalent radius of sodium is 166 pm, so the 3s shell nearly triple that of neon. That gap in sodium inside its 3s shell is much larger than 2s for the space required for the 3p shell to be inside 3s. 3p begins with aluminum.

Whether all orbitals are circular must be addressed in a practical atomic model. This is covered in section 5 Gather Electrons.

1.15 Certainty of an electron's position in an atom's cloud

The claim of an electron in Quantum mechanics having an uncertain position in an atom's electron cloud is also covered in section 5 Gather Electrons.

1.16 Relativity

Several of the author's books have addressed problems with relativity and space-time.

Space-time is the moving observer's reference frame. Relativity is background independent meaning it has no connection to physical coordinates. Everything is observed relative to the observer.

With that context, no other observer can observe or measure the universe with a common reference point as used by the special moving observer.

Einstein applied special rules for this special observer. Among them is a velocity limit at the velocity of light. This limitation had no justification at the time. Since then quasars have very high red shifts from protons traveling at high velocity toward the quasar, emitting the Lyman-alpha emission line having a z might greater than c.

Einstein applied a second rule having dire consequences. Only a gravitational field affected the path of this special observer. The 2 significant fundamental forces of electric and gravity are mutual and instantaneous. The mutual force is always present but is reduced by the inverse square of distance between participants.

Relativity presents an observer affected only by a gravitational field while this observer has no defined effect on other entities.
Nearly all matter in the universe is plasma when much of atomic matter is ionized. There are Birkelund current filaments found in many places, including from Sun to Earth, and these can consist of protons and electrons. When in plasma filaments, these oppositely charged particles flow together, bound by the magnetic field being generated.

Relativity is not a valid revision in explaining motion affected by the 3 fundamental forces, when both electric and magnetic are just ignored.

Einstein applied another wrong rule, where nothing (including "information") could travel faster than c. No instantaneous forces require time to travel, so the rule is meaningless.

Space-time is an attempt to manage external forces in an abstract manner, called curvature. Perhaps not by coincidence, quantum mechanics attempts to manage electrons in an abstract manner, called a wave function.

2 Fundamentals

There are several fundamental assumptions for an updated atomic model.

2.1 Electron
An electron is a fundamental particle having a measured mass and size.

Under certain conditions, it can flip its polarity from negative to positive, becoming a positron.

There is no other way for a positron to appear, except when an electron becomes one.

There is no known mechanism to create an electron or positron, other than one changing its polarity.

2.2 Proton

A proton is a fundamental particle having a measured mass and size.

A proton has severable notable behaviors.

a) Under certain, rare conditions, a proton can flip its polarity from positive to negative, becoming a antiproton. So far, this has been observed only in particle accelerators.

There is no other way for an antiproton to appear except when a proton becomes one.

There is no known mechanism to create a proton or antiproton, other than one changing its polarity.

Also, a proton having an attached electron is a neutron. A neutron is not a single coherent particle

PPPB justified a lower mass for a proton.

With its current value, the mass of the hydrogen atom is greater than the sum of an electron and proton. This should be impossible when there are no other particles present.

Using the measured mass of an electron and the known ratio between masses of electron and proton, the correct mass of a proton is easily calculated. Its current value (of too high) came from assumptions using carbon-12. This basis is a mistake because its nucleus has its protons distorted during fusion. It should be noted hydrogen was the first basis for atomic mass units. Oxygen was also tried, but was discarded due to its isotopes. Carbon is the current basis despite its known isotopes.

All analysis by this author uses this updated proton mass.

b) When a proton is fused against other protons in a nucleus, its volume is slightly reduced. This results in a tiny reduction in its measured mass. This is the cause of the anomaly called atomic mass defect.

c) While protons are in contact, the electrostatic force between those 2 positive charges flips its direction from repulsion to attraction. This change is not permanent. If the contact is disturbed, then the force reverts back to repulsion. Among the steps of radioactive decay are proton ejection and alpha particle ejection, which is the combination of 2 protons and 2 neutrons. These ejections occur when the equilibrium in the nucleus is disturbed.

d) When an electron is in contact with a proton, the combination is called a neutron because the combination has no net charge, or it is neutral.
The contact is maintained by the electrostatic force between opposing charges.

This change in direction is not permanent. If the contact is disturbed, then the two charges will separate. A neutron outside a nucleus will separate in a few minutes. A neutron inside a nucleus can be bound to other protons. Other neutrons bring their electrons so the nucleus gets a combination of positive and negative charges acting simultaneously. Stability for a long time requires equilibrium.

A mix lacking equilibrium can result in a step of radioactive decay.
For one example, a nucleus lacking enough neutrons is called "proton-rich" and one possible action is a proton ejection, to attempt stability with a loss of equilibrium. Another possible action is called an electron capture. One of the 2 inner electrons is captured by a proton in the nucleus. This increases by 1 the number of electrons in the nucleus. Depending on the number of excess protons, sometimes this capture can be enough for improved stability or even equilibrium. With this event, a proton became a neutron, though in reality the proton in the nucleus did not change in any way.

For another example, a nucleus having too many neutrons is called "neutron-rich" and one possible action is an electron's ejection from a neutron, to seek stability. With this event in the nucleus, a neutron became a proton, though in reality the neutron's proton in the nucleus did not change in any way.

In the Standard Model, the 3 quarks inside the neutron must somehow transform themselves, re-creating the original particle pair. This explanation is not acceptable in this updated atomic model. A neutron is always a pair.

In the Standard Model, the two merge becoming a single particle consisting of 3 quarks, just like a proton has 3 quarks.

This practical atomic model has no quarks and a neutron is always its 2 individual particles. This distinction, of 2 not 1, enables better explanations of some atomic behaviors.

The neutron explanation is one of the problems in the Standard Model.

3 Data Sets

The author compiled data into several sets to support this book's conclusions.

3.1 Caveat

There is a crucial limitation in analyzing the entire periodic table of elements.

One of the most important measurements of an atom is the radius of its outer valence ring of electrons. This ring is the primary participant in chemical bonds and its distance from the nucleus affects its electrons.

Currently, the valence radius is consistently available for only the first 86 elements, or Hydrogen through Radon. Element 87, Francium, and those heavier, do not consistently have this value. The reason must be related to these heavier elements being radioactive. The heaviest element having a stable isotope is Lead, having atomic weights of 206, 207, and 208 as stable.

There are isotopes having a measured mass for all 118 elements.

Therefore, some conclusions can be drawn for all 118, while other conclusions apply to a smaller set.

The charts present available data. Gaps result from no data.

3.2 Isotope Data

All the long life isotopes in the periodic table are analyzed for a mass defect, or the reduction in a proton's measured mass due to its compression during fusion, and radioactive decay.

Here, a long life isotope is one which is either stable or its half-life is long enough for using its nucleus for atomic analysis. For example, isotopes having a half-life less than 1 second are usually ignored for those few unstable elements having no isotopes lasting more a few seconds.
The author compiled data from all elements and their long life isotopes to compare each for their measured value against the sum of their components.
This reference file in .xls format is compressed in a .zip format for convenient distribution.

ZR-Isotopes.zip

Note: The main worksheet, using MS Excel, has over 1095 rows. Compressing that content into a smaller page is quite impractical.

Many of the 118 elements have several stable isotopes as well as multiple radioactive isotopes.

A readme text file in the zip helps explain the worksheet.

The first decay step for the radioactive decay of an isotope is noted for the author's research, as well as the daughter isotope remaining after the decay action.

The behaviors in a radioactive nucleus require a clear explanation in particle physics.

The process toward that goal begins with the analysis of this behavior among all the elements.

A stable isotope usually has consistent proton rich steps for isotopes with fewer neutrons, like beta plus or electron capture, and neutron rich decay steps for isotopes with more neutrons, like beta minus.
For example some stable isotopes are classified as "Observationally Stable" because isotopes, which the author shortens to "oS"

An oS isotope is among the mix of radioactive isotopes further from the stable isotopes which can have consecutive numbers of neutrons.

A work sheet with formulae expedites this analysis.

Each element has a mix of entries and calculations, such as one for the number of neutrons after entering protons and nucleons.

3.3 Element Data

All the elements in the periodic table are analyzed for their average pass per proton.

The respective radius values, like valence, are entered.

This reference file is Z-Elements in .xls format, for Microsoft Excel, and is compressed in a .zip format for convenient distribution.

Z-Elements-PAM.zip

Note: The main worksheet has over 118 rows, with 1 per element. Compressing that content into a smaller page is quite impractical.

The columns include the atomic mass details to calculate the average proton mass for each element.

When available, the atomic radius and the valence radius are in their columns.

When available, the element's energy values for the first ionization, second, third, to ninth ionizations are in their respective columns,

Several columns have a chart covering all the elements having that value.

The Wan der Waals radius is also entered, when available. That behavior and those values are not presented in this book. They were entered in case a need arose.

A readme text file in the zip helps explain the worksheet.

Several atomic behaviors require a better explanation in particle physics.

The process toward that goal begins with the analysis of each behavior among all the elements. Not all behaviors are covered; this is not a chemistry text book.

A work sheet with consistent formulae enables efficiency in this analysis. These data sets are used to justify the author's conclusions.

3.4 Building a Nucleus

A worksheet was created to compile data into a convenient format.
All the valid isotopes in the periodic table are analyzed for a stable nucleus. This done for all the known proton counts in atomic nuclei, from elements 1 to 118, for nucleon counts from 1 to 295.

This reference file in .xls format is compressed in a .zip format for convenient distribution.

Z-Nucleus-Builder.zip

Note: The main worksheet, using MS Excel, has about 295 rows, covering the relevant combinations of nucleons. Compressing that content into a smaller page is quite impractical.

The stability of a nucleus depends on the physical arrangement of the protons in combination with the arrangement of the electrons whose presence results in neutrons. This behavior of building a nucleus requires a better explanation in particle physics.

The process of developing an improved explanation begins with the analysis of this behavior among all the elements.

As noted earlier, a plasma pinch provides a substantial electromagnetic force capable of compressing charged particles into a nucleus.

3.5 Building an Electron Configuration

Later, a change in naming electron rings is proposed.
The author developed a VBA application for MS Excel to generate the electron configuration for all 118 elements. All the valid isotopes in the periodic table are analyzed for a stable nucleus for all the known proton counts in atomic nuclei, from elements 1 to 118, for nucleon counts from 1 to 295. The valence for each element is shown. This column of valences is presented in a chart.

This reference file in .xls format is compressed in a .zip format for convenient distribution.

Z-Populate-NewRings.zip

Note: The main worksheet, using MS Excel, has over 118 rows, covering all the currently known elements.

A readme text file in the zip helps explain the worksheet and operating its main macro.

The name of the file comes from its application of the new shell or ring names, replacing spdf.

4 Build Nucleus

This exercise describes the interaction between protons and electrons in a nucleus.

The most stable counts of protons in a nucleus are described. This is an investigation of the supposed strong force, which is claimed to hold a nucleus intact. As described in the author's PPPB, there is no such separate force. The protons and electrons in contact are mutually attracted by Coulomb's force.

There is no supposed weak force which is claimed to eject particles from a nucleus. When equilibrium is disturbed the Coulomb's force reverts to repulsion. The nucleus behaviors are driven by Coulomb's force and no other.

Force is required to fuse a proton with other protons, because the electrostatic force between positive charges repels them. However, if contact is attained, the repulsive force becomes attractive, until disturbed. Combinations of electrons and protons within the nucleus enhance its stability, because the protons are also bound to the electron(s).

When an electron is attached to a proton, the combination is called a neutron because the pair of attached particles exhibits no net charge.

The sequence of proton and electron combinations for a stable nucleus is not linear.

Oxygen is 16 protons with 8 electrons, so its atomic number is 8 for its 16 nucleons, and is identified as ^{16}O, where the number 16 is the number of nucleons in this isotope of Oxygen. The next stable element is Fluorine having 17 nucleons and 10 electrons, so its atomic number is 9 and is ^{19}F. A nucleus with 18 nucleons with 9 electrons, or ^{18}F, is not stable.

Some of the lighter elements have the number of electrons at half the number of nucleons. That works for ^{14}N, and ^{16}O, and others, but not F, and others.

The author plotted all the stable isotopes, or each with the longest life among alternatives, for all the nucleon counts from 1, for element 1 Hydrogen, to 295 for element 118 Oganesson.

The number of electrons accompanying each number of nucleons is charted.

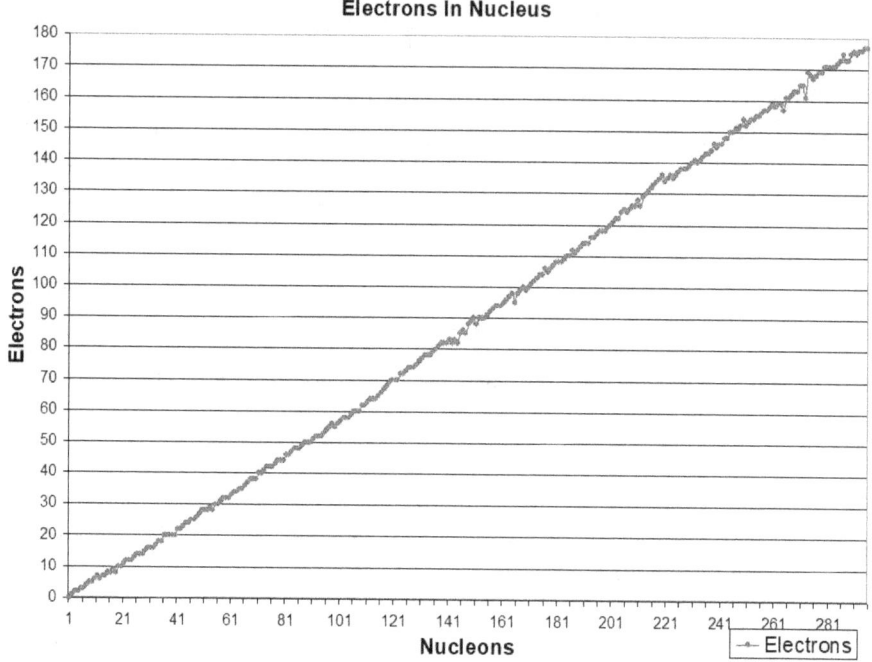

Electrons In Nucleus

The ratio is not linear, with a few anomalies at certain nucleon counts. Those will be explained later.

4.1 Protons in a nucleus

Protons get compressed during fusion causing their mass to decrease. Here is a chart showing this decrease.

The nominal mass of an atom is the sum of its protons and neutrons. An atom's measured mass includes the electrons orbiting the nucleus. In the nucleus, a neutron is a proton with an attached electron. The average mass per proton in an atom is calculated from the atom's measured mass, subtracting the mass of all the electrons, then dividing by the number of protons (atomic weight).

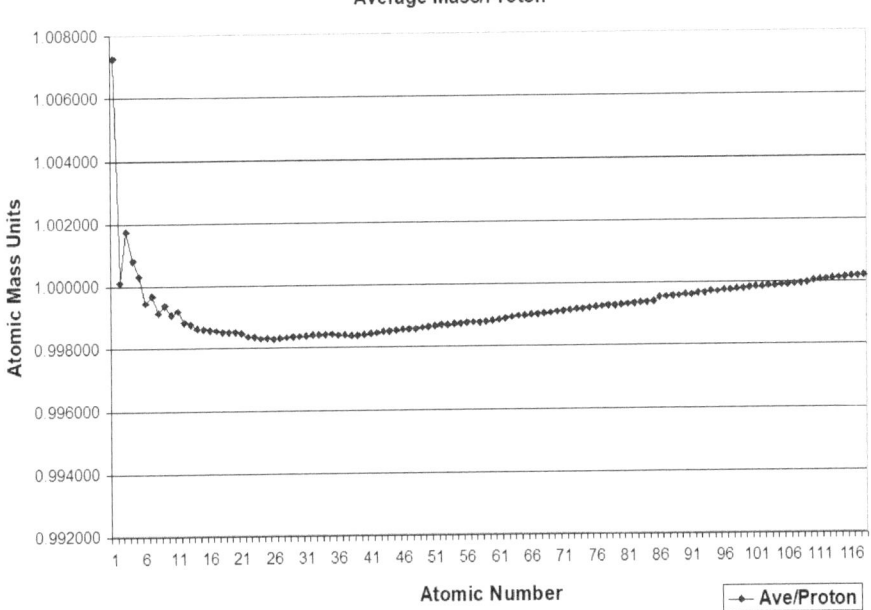

Average Mass/Proton

Hydrogen is atomic number 1. Its proton mass is Unchanged.

All the known isotopes which are not found in nature, like above 94 or Plutonium, are created by particle colliders using heavy nuclei. That mechanism probably results in inconsistent compression within the nucleus.

An atom consists of protons, neutrons, and electrons. Each neutron is the combination of a proton and electron. An atom has the same number of protons and electrons, so the non-ionized atom exhibits no net charge.

The above chart illustrates the cause of an atomic mass defect. The mass of a proton is more than 1 amu, but when measuring them compressed into a nucleus, a proton consistently measures at less than 1 amu.

Currently it is explained as a conversion of mass to nuclear binding energy. This wrong explanation for a mass defect and the author's better explanation were described in the author's previous book, Practical Particle Physics.

The process of fusing protons or neutrons into a nucleus compresses the proton resulting in a reduction in its measured mass. This mass defect is not nuclear binding energy as currently claimed. The Standard Model is wrong with its description of a nucleus.

4.2 Electrons in a nucleus.

Electrons in a nucleus with the protons enhance the stability of a nucleus. If there are too few electrons or too many for the number of protons, a step of radioactive decay occurs seeking a better combination.

Here is a chart presenting the combinations having the longest lives. The number of protons in the nucleus is the X-axis with its corresponding number of electrons plotted on the Y-axis.

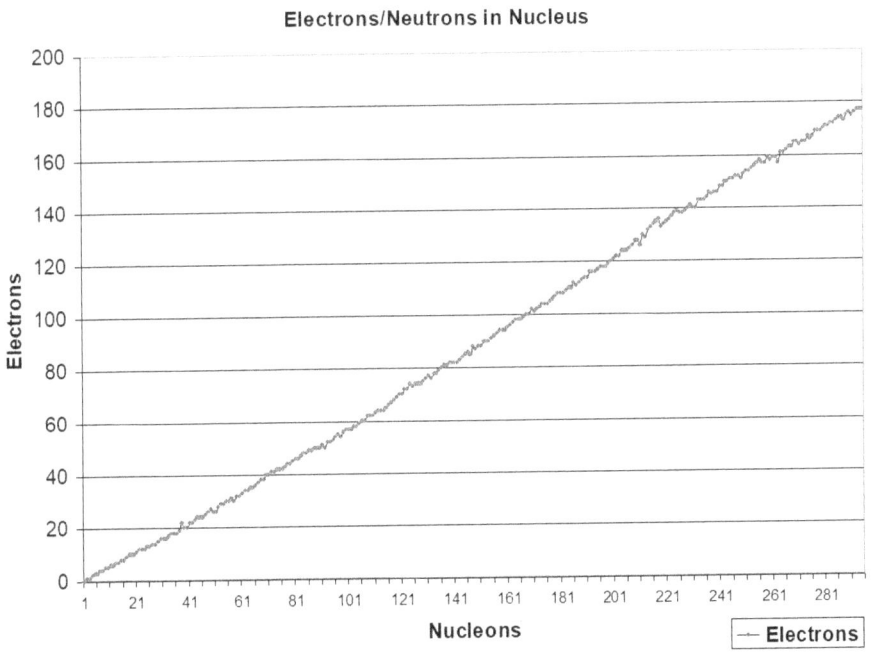

Heavy radioactive isotopes are at the far right.

The correct analysis of this behavior requires the nucleus topology or the packing arrangement of its nucleons. Being symmetrical or not can affect packing. Unfortunately, the topology might not be possible in most cases.

^{2}H or deuterium is stable with 2 protons and 1 electron. This is probably a triangle which is a symmetrical shape so its stability is expected.

^{3}He is stable with 3 protons and 1 electron. This is probably a triangle of the protons with the electron in contact will all 3 protons. This is also a symmetrical shape so its stability is expected.

There is no stable isotope with 5 nucleons. 4 electrons in the nucleus is 5H; 3 electrons is 5He; 2 electrons is 5Li; 1 electron is 5Be; none of these combinations are stable. If its shape for 5 spheres were a simple cube of 4 with the 5th in the center, then it is impossible for an electron to make contact with the proton in the center. The center proton is in a precarious equilibrium with its 4 adjacent protons while any electrons in the nucleus are attached on the outside of the 4 in the cube. Perhaps that topology is why the compact set of 5 is never stable.

The worksheet has a column indicating an odd number of protons in the nucleus. Elements having an even number of protons often have more stable isotopes than those elements having an odd number.

4.3 Size of an atom

Most, but not all, have a measured atomic radius in pm.
None of the noble gas elements have a public value.

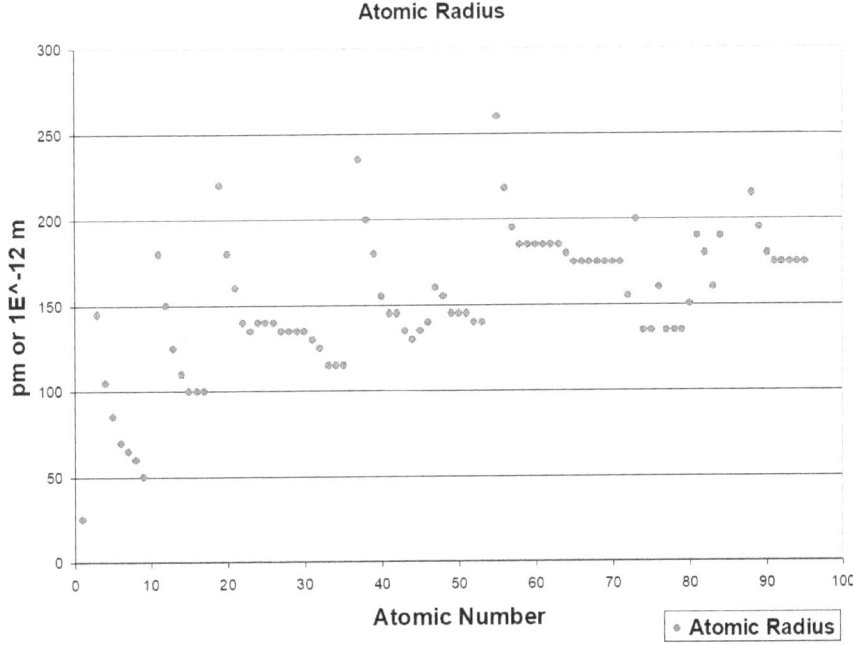

Atomic Radius

The peaks are the group 1 elements where the b ring begins each cycle. As more protons are added, while moving through the periodic table, the atomic radius decreases due to increasing protons in nucleus.

The flat sequences where consecutive elements have nearly the same radius occur when filling the d ring, to 10, or the f ring, to 14.

4.4 Covalent Radius

Most, but not all, elements have a measured atomic covalent radius in pm, which is the last shell in the sequence.

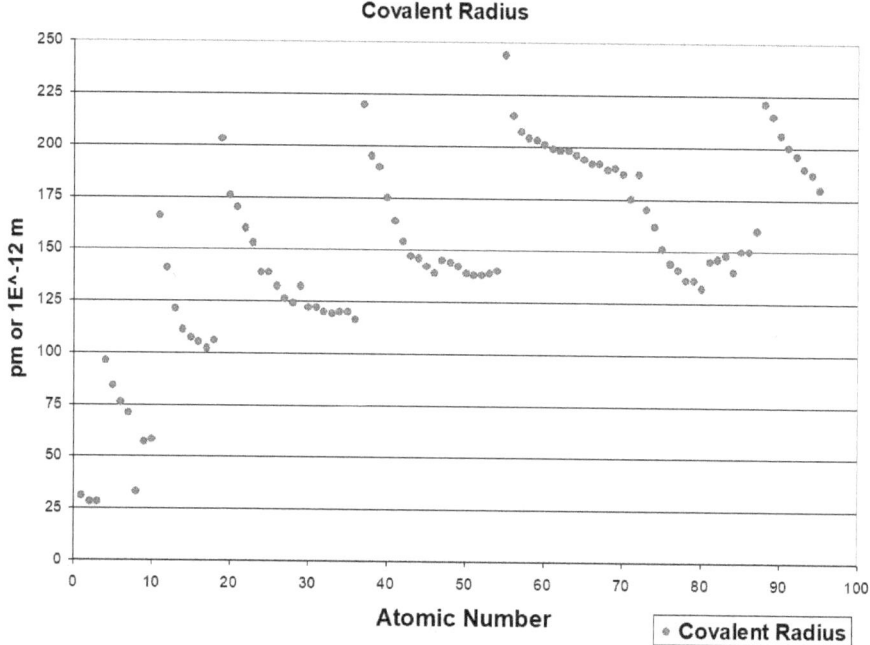

The size roughly decreases as the number of protons is increasing toward the right in a row of the periodic table.

This is the result of increasing the positive charges which increases the mutual electrostatic force, as implied by Coulomb's law.

This book will consistently use covalent radius and not atomic radius because a covalent radius value is available for all elements.

Though these values of atomic or valence radius have only 3 significant digits, there is never a stated margin of error for any values.

If electrons were truly moving based on probabilities, then the measured values should vary based on electron actual positions at the moment. The fact that these values have no deviations confirms electrons are always in their expected positions. The supposed electron uncertainty principle does not exist; if it did, then there must be some evidence for it. There is none.

4.5 Adding to a Nucleus

There are 2 known mechanisms for adding particles to a nucleus: fusion and transmutation.

4.5.1 Fusion

Fusion is assumed to occur in the cores of stars, including our Sun. Extreme pressure and temperature are required to overcome the electrostatic force of repulsion between protons.

The author's book Cosmology Transition provided explanations and references to a solar model not based on a gaseous sphere of plasma powered by fusion. This alternative is based on condensed matter, specifically liquid metallic hydrogen, which is a lattice of protons maintained by loose electrons.

4.5.2 Plasma Pinch

This mechanism is probably responsible for the creation of both spherical stars and spherical planets

Excerpt from Wikipedia:

A pinch is the compression of an electrically conducting filament by magnetic forces, or a device that does such. The conductor is usually a plasma, but could also be a solid or liquid metal. Pinches were the first type of device used for controlled nuclear fusion.

The phenomenon may also be referred to as a Bennett pinch (after Willard Harrison Bennett), electromagnetic pinch, magnetic pinch, pinch effect or plasma pinch. Pinches occur naturally in electrical discharges such as lightning bolts, the aurora, current sheets, and solar flares.

(Excerpt end)

Observation:

The excerpt notes this mechanism was used for the first attempt at controlled fusion. Duplicating the impossible conditions claimed to be in the Sun's core is also impossible.

4.5.3 Transmutation

Radioactive elements having short half lives have been detected on a distant star. The only reasonable explanation is they are formed on the star's surface, where they are being detected. There is no other reasonable mechanism to explain their presence.

As the video mentioned next notes, some astronomers have failed to find an explanation so they offer the preposterous conjecture of aliens putting them there. The heavy elements include Einsteinium which was rarely observed outside of particle colliders due to its short half-life.

The title of the YouTube video by Sky Scholar is:

Przybylski's Star & The Total Denial of Reality - Aliens, Undiscovered Elements, and more!

Link is in References.

Transmutation in natural, biological processes has been known for many years.

The title of the YouTube video by See the Pattern is:

Biological Transmutation of Elements

Several experiments around the world have found elements which were not present at the start. These have been called cold fusion.

The title of the YouTube video by See the Pattern is:

Experimental Transmutation of Elements

Link to the video is in References.

LENR or Low Energy Nuclear Reactions is a topic for current research. The topic is out of scope for this book and the Practical Atomic Model does not rely on specific mechanisms for changing a nucleus beyond the known steps of radioactive decay, which is the topic of Section 7.

4.6 Even and odd Nucleon Counts

A nucleon is usually a proton or neutron. Because a neutron is just a proton having an attached electron, in this section, the term nucleons refers to the atomic weight or the number of protons and neutrons counted in the nucleus. There is a distinction between electrons orbiting the nucleus and those attached to protons in the nucleus.

Section 4.5 describes electrons attached to protons in the nucleus.

This number matches the number of neutrons.

However, the behavior is easier to describe as attached electrons rather than the number of neutrons.

The current, typical description of a nucleus is a number of protons and neutrons.

The new description of a nucleus is a number of protons and the number of attached electrons.

The number of electrons required for a stable nucleus varies on whether the number of nucleons is odd or even.
The first 5 counts are simple. The number of electrons is half the number of nucleons with the result rounded down.

The resulting series:
1 gets none; 2 or 3 get 1 attached electron.
1 proton in a nucleus needs no electrons because ^{1}H or hydrogen is stable. 2 protons in the nucleus need 1 electron and is ^{2}H or deuterium which is stable.
3 protons in the nucleus need 1 attached electron and results in ^{3}He which is stable.

All of the nucleon counts from 6 to 17 follow a simple rule:

a) If the count is even, then the number of attached electrons is half the number of nucleons rounded down.

b) If the count is odd, then the number of attached electrons is half the number of nucleons rounded down and adding 1.

An example of an even count is 4. Halving 4 and rounding down results in 2 electrons, which is ^{4}He.

An example of an odd count is 7. Halving 7, rounding down, and adding 1, results in 4 electrons, which is ^{7}Li.

This rule does not apply to 5 nucleons. There is no stable isotope having 5 nucleons. This is probably because of the topology of 5 packed spheres. It is impossible for an electron to attach to the internal proton surrounded by 4 protons. That is the most likely densest packing of 5.

This rule was applied to predict the number of electrons for all the nucleon counts from 1 to 252 and compare that number against that of the isotope being stable, or having the longest half-life.

The Nucleus-Builder.xls has a worksheet titled Nucleus and a macro which predicts the required electron count and calculates the extra electrons being fused into the nucleus to achieve its stability or longevity.

Here is a chart of these extra electrons beyond the simple even and odd rule, in relation to the number of nucleons.

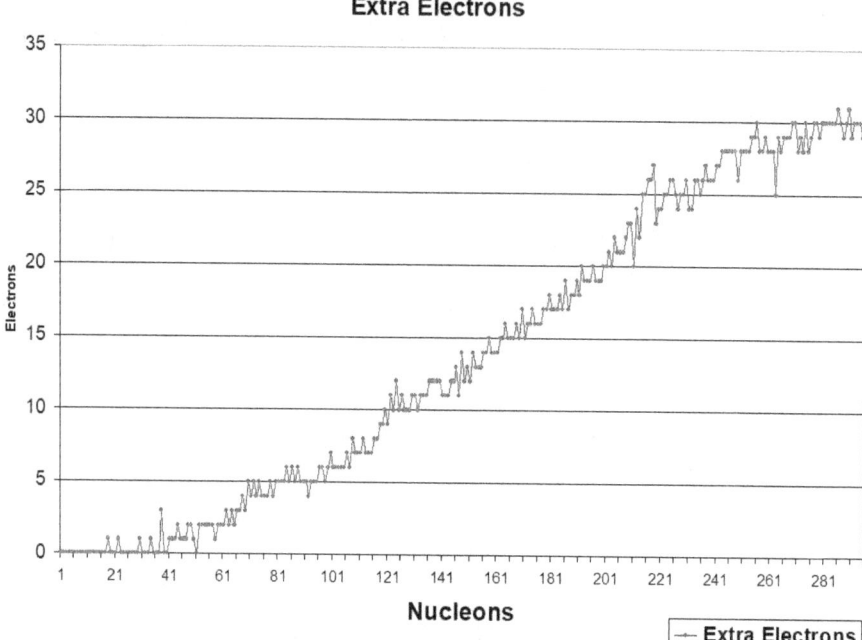

Extra Electrons

The lower left shows those nucleon counts less than 18 follow the simple even (half) and odd (half + 1). Frequently, consecutive even and odd counts of nucleons are different elements, where the extra electrons accompanied a change in the number of protons after subtracting the number of neutrons.

The first extra electron is at 18 nucleons. The rule for even nucleons is half for 9 electrons and results in a nucleus of 9p + 9n but ^{18}F is unstable. One extra electron results in ^{18}O with 8p +10n and is stable.

The second extra electron is at 22 nucleons. The rule for even is half (for 11 electrons which results in a nucleus of 11p + 11n but ^{22}Na is unstable. One extra electron is ^{22}Ne with 10p +12n and is stable.

The third extra electron is at 30 nucleons. The rule for even is half, for 15 electrons which results in a nucleus of 15p + 15n but ^{30}P is unstable. One extra electron is ^{30}Si with 14p +16n and is stable.

The fourth extra electron is at 34 nucleons. The rule for even is half for 17 electrons which results in a nucleus of 17p + 15n but ^{34}Cl is unstable. One extra electron is ^{34}S with 16p +18n and is stable.

36 Nucleons follows the rule because ^{36}S with 18p + 18n is stable.
The next count needing extra electrons is at 38 nucleons. The rule for even is half for 19 electrons which results in a nucleus of 19p + 19n but ^{38}K is unstable. 3 extra electrons results in ^{348}Ar with 16p +18n and is stable.

40 Nucleons follows the rule because ^{40}Ca with 20p + 20n is stable.

The first extra electron with an odd number of nucleons is at 41 nucleons. The rule for odd is half+1 which is 21 electrons and results in a nucleus of 20p + 21n but ^{41}Ca is unstable. One extra electron is ^{41}K with 19p +22n and is stable.

The second extra electron with an odd number of nucleons is at 43 nucleons. The rule for odd is half+1 which is 22 electrons and results in a nucleus of 21p + 22n but ^{43}Sc is unstable. One extra electron is ^{43}Ca with 20p +23n and is stable.

These combinations are presented in the above chart. Rather than describing all 252 nucleon combinations in this book, which could require nearly 252 pages, the author recommends to those having access to spreadsheet software open the spreadsheet file identified in the Section 3 Data Sets.

5 Gather Electrons

The orbitals described in the Standard Model do not aid in understanding an atom.

5.1 Current Naming of orbitals.

An excerpt from Fundamentals:

" This gives the following order for filling the orbitals:

1s, 2s, 2p, 3s, 3p, 4s, 3d, 4p, 5s, 4d, 5p, 6s, 4f, 5d, 6p, 7s, 5f, 6d, 7p, (8s, 5g, 6f, 7d, 8p, and 9s). "

Element 118 has 19 steps through 7p. There are no known elements beyond 118 and beyond 7p in that specified shell sequence.

Element 1 or Hydrogen has a single step.

By putting Hydrogen at the first step in this pattern has a first step of 1.
This sequence is:
1,2,6,2,6,2,10,6,2,10,6,2,14,10,6,2,14,10,6.

This image presents that pattern.

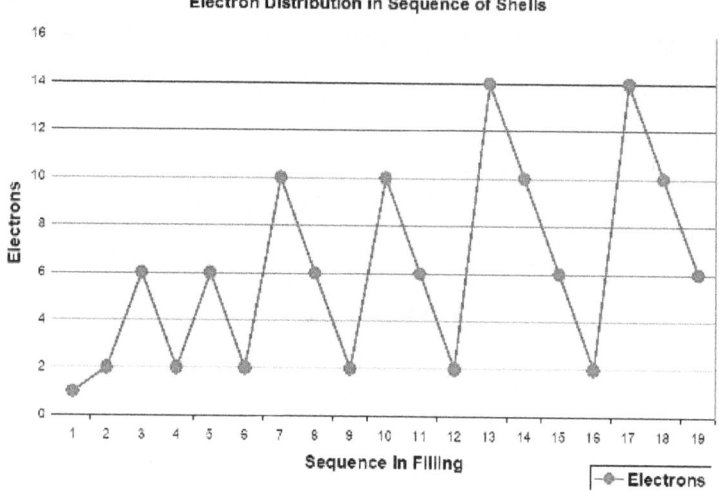

There is clearly a pattern.

An orbital having 2 electrons is at the beginning of each cycle.

The current orbital naming convention is not helpful. The shell 4f is after 5p and 6s. A number is often expected to indicate a state having consecutive values. The naming of the f shells fails that expectation.
The number preceding "s" could be called a cycle number. Over the course of the periodic table, the maximum number of electrons per shell progresses from 2, to 6, to 8, to 10, to 14, with the maximum repeating in consecutive cycles or sometimes increasing in alternating cycles.

5.2 Replacing the word Orbital

The word orbital is not quite right for an atom.
The word shell has been used because of its spherical connotation. Orbits in our solar system are actually ellipses, not circles, because the motion is around the system's center of gravity. Electrons do not move like planets around the Sun by the force of gravity. Electrons move around the nucleus by the electric force between charges. Increasing the mass in a nucleus by adding neutrons does not change the electron configuration, so gravity is not involved within the atom and its orbiting electrons. Electrons are found in concentric rings.

Henceforth, the word ring will be used replacing the confusing words orbital or shell.

It is the author's opinion the word ring is better for an atom. The word shell implies a structure. A number of objects arranged as a circle around a center is called a ring, not a shell.

A positively charged nucleus attracts a ring of electrons around it. These rings hold more electrons as the positive charge in the nucleus increases.

There are several distances where equilibrium of electrostatic forces is achieved. These electrons are not moving in strange orbits driven by probabilities, while in equilibrium with the nucleus and other electrons.

5.3 Shapes of Rings

As noted earlier, atomic rings have 2, 6, 10, or 14 electrons.

The following represent those symmetrical patterns, where an electron is at each intersection around the circumference.

There are 3 pairs of opposing electrons.

6 points

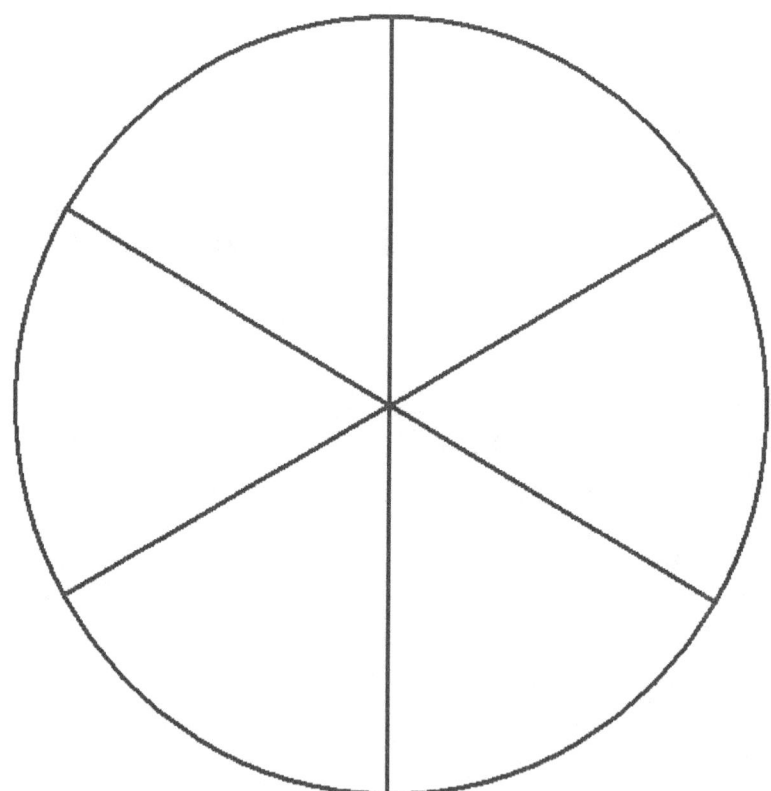

This symmetry enables equilibrium.

When filling a ring of 6, usually, the 2 electrons from the existing outer b ring are merged with the x ring, resulting in the x ring having a capacity of 8 electrons.

As a reminder, the ring has this symmetry.

8 points

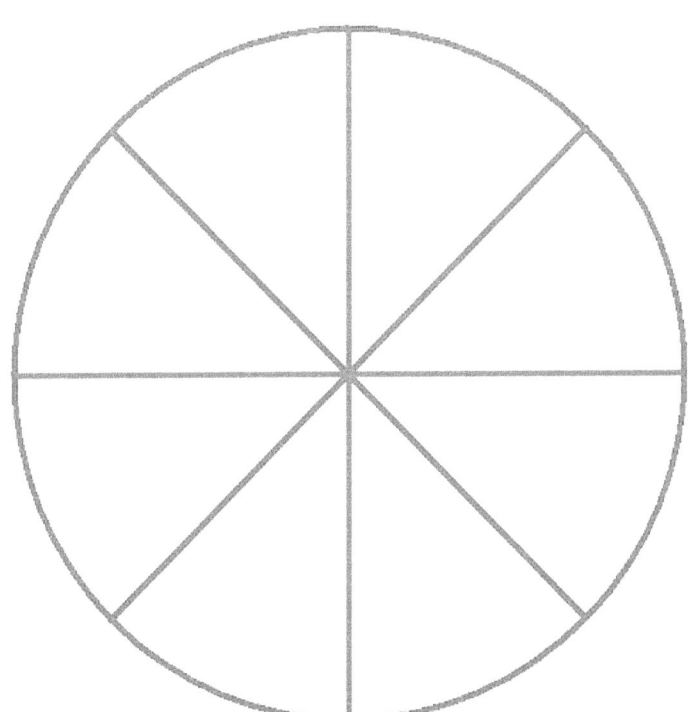

There are 4 pairs of opposing electrons.

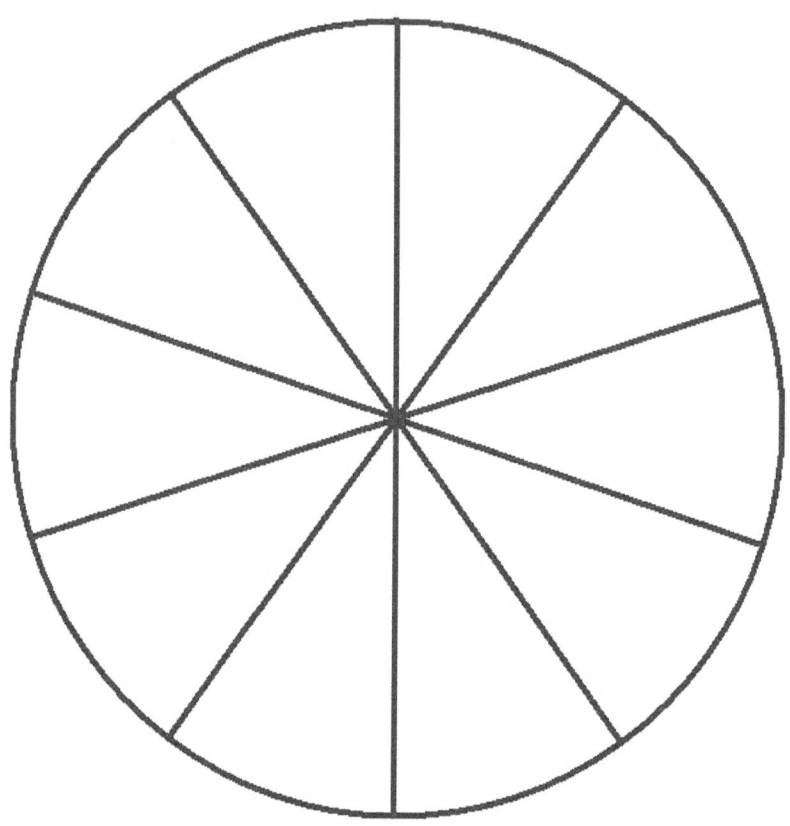

There are 5 pairs of opposing electrons.
This symmetry enables equilibrium.

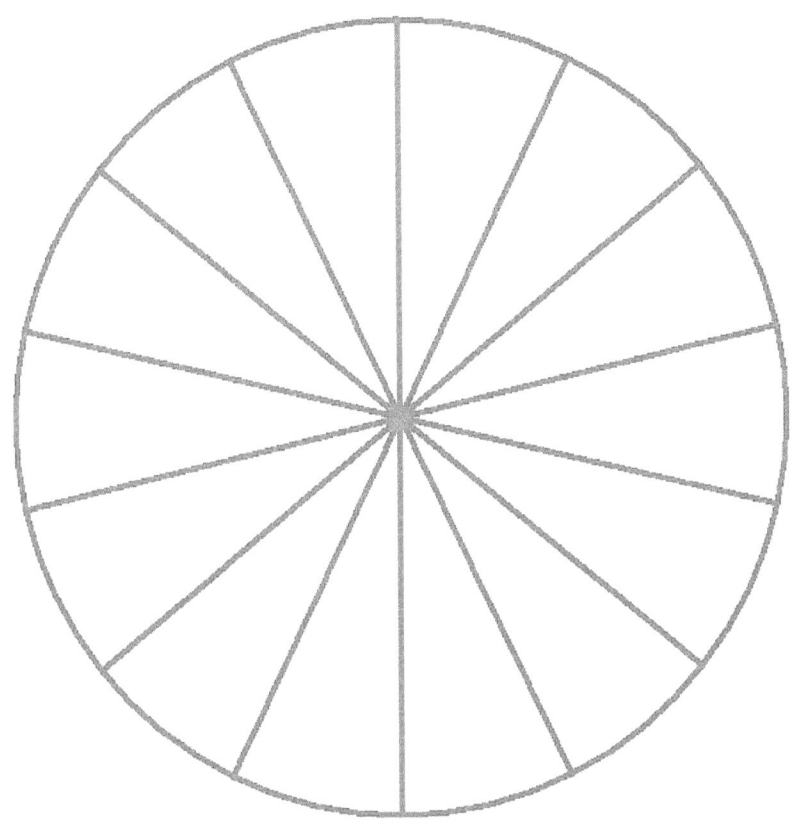

There are 7 pairs of opposing electrons.
This symmetry enables equilibrium.

5.4 Change the naming of the rings

The naming should help describe the configuration. The current shell names are a jumble of non-consecutive numbers with 4 letters indicating nothing of ithe meaning of the combination.

Part of the current shell sequence has: "6s, 4f, 5d," with 4 after 6 and before 5. This confusing convention of names can be improved.

The author proposes a new naming convention. The numbers become consecutive because the electron rings are filled in a particular pattern. The 4 recommended letters directly imply the maximum number of electrons in its ring.
.
Fortunately for a transition, the letters d and f are maintained. Several words imply the number 10 but the prefix dec- is often used, including the metric naming system. This is the basis for maintaining d for 10.

The letter f for fourteen is intuitive in English.

14 in Spanish is catorce but the letter c, being the 3rd letter and implying 3, is a poor choice.

14 in French is quatorze but the letter q, from the quad-prefix and implying 4, is also a poor choice.

The letter f is maintained for 14.

The letters s and p could not be related to their number, so these become b and x, respectively. Several words suggest the number 2 but the prefix bi- is often used. This is the basis for selecting b. The letter x is found in both six and hexagon.

When an atom is filling its x ring in combination with the outer b ring, the b+x ring changes capacity from 2 or 6 to a capacity of 8 by adding the 2 electrons from the previous b ring to those in the x ring being filled.

Unfortunately, the letter x is sometimes associated with an unknown value. However In this naming, the letter x will imply 6 for the sequence where x ring has merged with the outer b ring, the letter x means "extra" because its capacity becomes 8 or it can hold 2 extra electrons.

For clarity, the x and b rings have their distinct counts of electrons. When presenting the number of valence electrons are summed between the two rings, but the 2 separately defined rings actually coincide in 1 ring. The number of electrons in this combined ring will range from 1 or 2, the normal range for b ring, while the x ring is empty. When the X ring begins filling, the b ring for that cycle will be filled. As a result, the range of electrons in this outer ring, which is providing the valence electrons from the combination of an x ring and b ring, will range from 2 to 8.
A ring for 6 is present in every atom having more than 4 electrons so its letter of p could warrant "principle" (one current justification for the letter p) but the author recommends consistency by relating each letter to its number of electrons in the ring.

5.5 Sequence of Gathering Electrons

The sequence that the atoms collect their electrons is clear with the combination of the atomic radius and the periodic table.

The radius of each electron's configuration has been measured in pm.

Ring cycle 1

1b=fill to 2; H to He; 1:2

H covalent radius = 3; 1b1
He covalent radius = 28; 1b2

Ring cycle 2

Ring 2b begins at roughly the same radius as the completed 1b ring at the end of cycle 1. No margin of error vale is provided for these radius values measured in 10^{-12}m or pm.

2b=fill to 2; Li,Be; 3:4
2x=fill to 6; B to Ne; 5:10; 2x+2b is valence 3 to 8

The radius of each electron configuration has been measured in pm.

3 Li = 28; 2b1
4 Be = 96; 2x2
5 B = 84; 2x1+2b2
6 C = 76; 2x2+2b2
7 N = 71; 2x3+2b2
8 O = 33; 2x4+2b2
9 F = 57; 2x5+2b2
10 Ne = 58; 2x6+2b2

The first 3 elements from H to Li are compact.

The next 4 are larger.
The 8th, O, is compact
9 & 10 are at a middle size.

Filling the 2x ring somewhat pushes out the outer ring of the combined 2b+2x rings. Li and O are smaller than the others in the 2x sequence.

Perhaps by its symmetry, Oxygen with 1b2, 2b2+2x4 (a symmetrical outer ring of 6 electrons around the inner ring of 2 electrons) is more compact than the other elements in this sequence of filling the 2b+2x ring.

Ring cycle 3

Ring 3b begins at more than double the radius of the completed 2x+2b ring at the end of cycle 2. The increase is from 58 to 166.

Cycle 3 matches the sequence of cycle 2 with the insertion of an x ring inside of the outer 3b ring.

3b=fill to 2; Na; Mg; 11:12; 3b is outer ring for valence

3x=fill to 6; Al to Ar; 13:18; 3x+3b is valence 3 to 8

The radius of each electron configuration has been measured in pm.

11 Na = 166, 3b1
12 Mg = 141, 3b2

Na is 1b2, 2b2+2x2 in the inside, with 3b1 on outside.
Mg with 1b2, 2b2+2x2, with 3b2 and is slightly smaller than Na.

The radius of each electron configuration has been measured in pm.

13 Al = 121, 3x1+3b2
14 Si = 111, 3x2+3b2
15 P = 107, 3x3+3b2
16 S = 105 ,3x4+3b2
17 Cl = 102 , 3x5+3b2
18 Ar = 106, 3x6+3b2

The atoms decreased in size as both protons an electrons increased by the same increments.

The attraction from the core overcomes the repulsion between the electron rings.
The electron counts in adjacent rings do not match to develop a synchronism.

Ring Cycle 4

Ring 4b begins at nearly double the radius of the completed 3x+3b ring at the end of cycle 3. The increase is from 106 to 203.

Cycle 4 requires the insertion of a d ring inside of the outer b ring, similar to an x-before-b combination in cycle 3 but cycle 4 inserts a d ring before the subsequent x ring.

4b=fill to 2; K, Ca;19:20; 4b is outer ring for valence
4d is filled inside 4b
4d=fill to 10; Sc to Zn; 21:30; 4b is valence
4x=fill to 6; Ga to Kr;31:36; 4x+4b is valence 3 to 8

The radius of each electron configuration has been measured in pm.

19 K= 203, 4b1
20 Ca = 176, 4b2
21 Sc = 170 , 4b2
22 Ti = 160 ,4b2
23 V = 153 , 4b2
24 Cr = 139 , 4b1
25 Mn = 139 , 4b2
26 Fe = 132 ,4b2
27 Co = 126, 4b2
28 Ni = 124, 4b2
29 Cu = 132, 4b1
30 Zn = 122, 4b2
31 Ga = 122, 4x1+4b2
32 Ge = 1204x2+4b2
33 As = 1194x3+4b2
34 Se = 1204x4+4b2
35 Br = 1204x5+4b2

36 Kr = 1164x6+4b2

Ring Cycle 5

Ring 5b begins at more than nearly double the radius of the completed x+4b ring at the end of cycle 4. The increase is from116 to 220.

Cycle 5 matches the sequence of cycle 4, with the insertion of a d ring and an x ring inside of the outer b ring.
5b=fill to 2; Rb; Sr, 37, 38; 5b is outer ring for valence
5d is filled inside 5b

5d=fill to 10; Y to Cd; 39:48; 5b is valence
5x=fill to 6; In to Xe;49:54; 5x+5b is valence 3 to 8

The radius of each electron configuration has been measured in pm.
37 Rb = 220, 5b1
38 Sr = 195, 5b2
39 Y = 190, 5b2
40 Zr = 175, 5b2
41 Nb = 164, 5b2
42 Mo = 154, 5b1
43 Tc = 147, 5b2
44 Ru = 146, 5b2
45 Rh = 142, 5b1
46 Pd = 139, 5b2
47 Ag = 145, 5b1
48 Cd = 144, 5b2
49 In = 142, 5x1+5b2
50 Sn = 139, 5x2+5b2
51 Sb = 138, 5x3+5b2
52 Te = 138, 5x4+5b2
53 I = 139, 5x5+5b2

54 Xe = 140, 5x6+5b2

Ring Cycle 6

Ring 6b begins at over 70% greater the radius of the completed 5x+5b ring at the end of cycle 5. The increase is from 140 to 244.

Cycle 6 has 1 more ring than the sequence of cycle 5, with the insertion of an f ring followed by the sequence in cycle 5 of a d ring and an x ring inside of the outer b ring.

6b=fill to 2; Cs, Ba; 55:56; 6b is outer ring for valence
6f is filled inside 6b
6f=fill to 14; La to Yb; 57:70; with 6b as valence

6d is filled inside 6b
6d=fill to 10; Lu to Hg; 71:80; with 6b as valence

6x=fill to 6; Tl to Rn; 81:86; 6x+6b is valence 3 to 8

The radius of each electron configuration has been measured in pm.

55 Cs = 244, 6b1
56 Ba = 215, 6b2
57 La = 207, 6b2
58 Ce = 204, 6b2
59 Pr = 203, 6b2
60 Nd = 201, 6b2
61 Pm = 199, 6b2
62 Sm = 198, 6b2
63 Eu = 198, 6b2
64 Gd = 196, 6b2

65 Tb = 194, 6b2
66 Dy = 192, 6b2
67 Ho = 192, 6b2
68 Er = 189, 6b2
69 Tm = 190, 6b2
70 Yb = 187, 6b2
71 Lu = 175, 6b2
72 Hf = 187, 6b2
73 Ta = 170, 6b2
74 W = 162, 6b2
75 Re = 151, 6b2
76 Os = 144, 6b2
77 Ir = 141, 6b2
78 Pt = 136, 6b1
79 Au = 136, 6b1
80 Hg = 132, 6b2
81 Tl = 145, 6x1+6b2
82 Pb = 146, 6x2+6b2
83 Bi = 148, 6x3+6b2
84 Po = 140, 6x4+6b2
85 At = 150, 6x5+6b2
86 Rn = 150, 6x6+6b2

Ring Cycle 7

All the elements in cycle 7 do not have a public covalent radius value.

Ring 7b must begin with an increase in radius from the completed 6x+6b ring at the end of cycle 6. That radius was 150 pm.

7b1 is at 160, is slightly larger. 7b2 is wider at 221. As the 7f ring fills inside the 7b ring, the valence radius continues around 200. With an increasing atomic number or a higher + charge in the nucleus, the valence ring radius also decreases.

Cycle 7 repeats the ring sequence of cycle 6, with the insertion of an f ring, a d ring, and an x ring inside of the outer b ring.

7b=fill to 2; Fr, Ra; 87:88; 7b is outer ring for valence
both 7f and 7d are inside 7b
7f is filled inside 7b, inside the upcoming 7d and the subsequent 7x; x then b ends the cycle.

7f=fill to 14; Ac to No; 89:102, with 7b as valence
7d is filled outside 7f but inside 7b
7d=fill to 10; Lr to Cn; 103:112; with 7b as valence
7x=fill to 6; Nh to Og; 113:118; 7x+7b is valence 3 to 8

The valence radius of some electron configurations been measured in pm.

87 Fr = 160, 7b1
88 Ra = 221, 7b2
89 Ac = 215, 7d1+7b2

90 Th = 206, 7d2+7b2
91 Pa = 200, 7d1+7b2
92 U = 196, 7d1+7b2
93 Np = 190, 7d1+7b2
94 Pu = 187, 7b2
95 Am = 180, 7b2
96 Cm = 169, 7b2

Elements 97 through 118 have no value for valence radius.

Note:
Elements 89 through 97 have an anomaly where the electrons filling the 7f ring are actually moved to the 7d ring, which normally fills after 7f.
The 7d electrons are present in the some of the valence counts.

Element 102 has its 7f ring filled with an empty 7d ring.
Element 113 begins filling its 7d ring after its 7f ring is filled.

5.6 Algorithm Demonstration

The author created a spreadsheet, using Microsoft Excel 2008, to demonstrate the predictability of the electron rings for every element, from 1 to 118.

This file was mentioned in section 3 Data Sets.

atomic weight	Sym.	1b	2b	2x	3b	3x	4b	4d	4x	5b	5d	5x	6b	6f	6d	6x	7b	7f	7d	7x	Elec total	Last Ring	Valence
1 H	H	1																			1	1b1	1
4 He	He	2																			2	1b2	2
7 Li	Li	2	1																		3	2b1	1
9 Be	Be	2	2																		4	2b2	2
11 B	B	2	2	1																	5	2x1	1
12 C	C	2	2	2																	6	2x2	2
14 N	N	2	2	3																	7	2x3	3
16 O	O	2	2	4																	8	2x4	4
19 F	F	2	2	5																	9	2x5	5
20 Ne	Ne	2	2	6																	10	2x6	6
23 Na	Na	2	2	6	1																11	3b1	1
24 Mg	Mg	2	2	6	2																12	3b2	2
27 Al	Al	2	2	6	2	1															13	3x1	1
28 Si	Si	2	2	6	2	2															14	3x2	2
31 P	P	2	2	6	2	3															15	3x3	3
32 S	S	2	2	6	2	4															16	3x4	4
35 Cl	Cl	2	2	6	2	5															17	3x5	5
40 Ar	Ar	2	2	6	2	6															18	3x6	6
39 K	K	2	2	6	2	6	1														19	4b1	1
40 Ca	Ca	2	2	6	2	6	2														20	4b2	2
45 Sc	Sc	2	2	6	2	6	2	1													21	4d1	1
48 Ti	Ti	2	2	6	2	6	2	2													22	4d2	2
51 V	V	2	2	6	2	6	2	3													23	4d3	3
52 Cr	Cr	2	2	6	2	6	1	5													24	4d5	5
55 Mn	Mn	2	2	6	2	6	2	5													25	4d5	5
56 Fe	Fe	2	2	6	2	6	2	6													26	4d6	6
59 Co	Co	2	2	6	2	6	2	7													27	4d7	7
59 Ni	Ni	2	2	6	2	6	2	8													28	4d8	8
64 Cu	Cu	2	2	6	2	6	1	10													29	4d10	10
65 Zn	Zn	2	2	6	2	6	2	10													30	4d10	10
70 Ga	Ga	2	2	6	2	6	2	10	1												31	4x1	1
73 Ge	Ge	2	2	6	2	6	2	10	2												32	4x2	2
75 As	As	2	2	6	2	6	2	10	3												33	4x3	3
79 Se	Se	2	2	6	2	6	2	10	4												34	4x4	4
80 Br	Br	2	2	6	2	6	2	10	5												35	4x5	5
84 Kr	Kr	2	2	6	2	6	2	10	6												36	4x6	6
85 Rb	Rb	2	2	6	2	6	2	10	6	1											37	5b1	1
88 Sr	Sr	2	2	6	2	6	2	10	6	2											38	5b2	2
89 Y	Y	2	2	6	2	6	2	10	6	2	1										39	5d1	1
91 Zr	Zr	2	2	6	2	6	2	10	6	2	2										40	5d2	2
93 Nb	Nb	2	2	6	2	6	2	10	6	2	3										41	5d3	3
96 Mo	Mo	2	2	6	2	6	2	10	6	1	5										42	5d5	5

Each ring has a column. The far right column has the valence electrons.
The order of the rings in the columns matches their increasing radius from the nucleus. This is not an alphabetical order.

5.7 Anomalies of an electron moved between rings

Elements Vanadium, Chromium, to Manganese are atomic numbers 23 to 25.
Chromium has 1 electron in its 4s orbital but the adjacent elements have 2.

The expectation for consecutive elements is the s orbital will be consistent after it fills to 2.

Explaining these anomalies is easier with better terminology and understanding the relationship between consecutive orbitals.

The s orbital is described as circular but p, d, f are elliptical or have lobes.

Several elements have anomalies where an electron has moved to an adjacent ring.

One example is Chromium or element 24. Its expected pattern is 4d4 + 4s2, but it has 4d5 + 4s1.

There are others between d and s.
There are several exchanges between f and d shells also.

Exchanges like this are reasonable with circular orbits, but electrons changing to different orbital paths could be awkward.

The author found no list of these anomalies in the known electron configurations. If one is published, it was not found.

The 20 elements with these atomic numbers have an anomaly where their electron configuration does not match that expected:
24, 29, 41, 42, 44, 45, 46, 47, 64, 78, 79, 89, 90, 91, 92, 93, 96, 97, 103, 111.

1 or 2 electrons are in an adjacent ring rather than where expected.

All of these anomalies are identified with their element in the section 8 Periodic Table.

5.8 Valence Electrons

Neither the current atomic model nor the Bohr model explains the mechanism for the valence electrons. By naming the rings by their distance from the nucleus, the outer-shell, or outer-ring, electrons are identified easily.

This chart presents that behavior for all the elements.

Valence Electrons

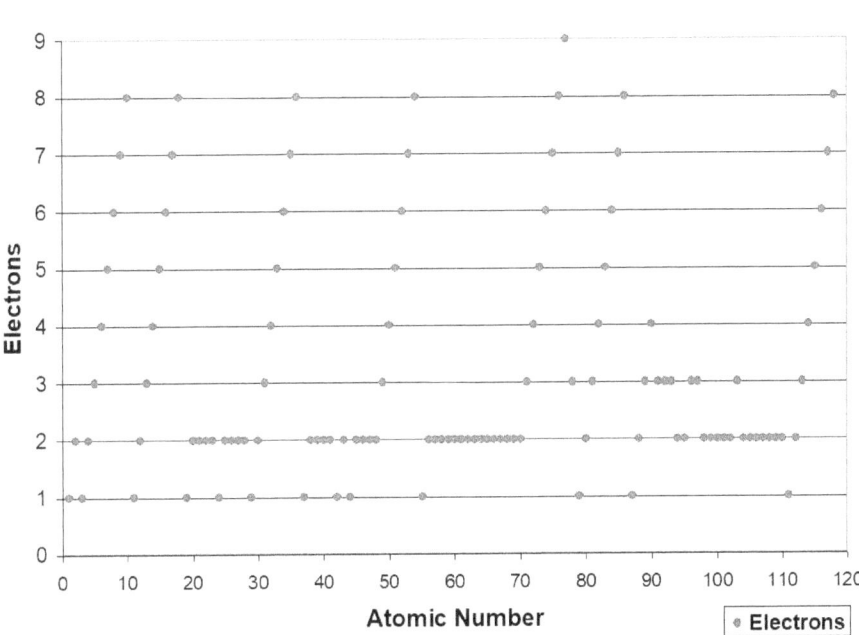

Iridium, or element 77, has 9 valence electrons, the highest in the chart. When the b ring begins it has 1 then 2. When the x+b ring is filling, it is incrementing from 3 to 8. While a d or f ring is filling, the b is the outer ring so the valence remains at 2 from the b ring, unless the filling ring interacts with the b ring. Sometimes, an electron will be exchanged between two rings.

5.9 New naming convention for electron rings

The following table defines a transition to the new ring names.

Old	New	type	cycle
1s	1b	binary	1st cycle
2s	2b	binary	2nd cycle
2p	2x	six	2nd cycle
3s	3b	binary	3rd cycle
3p	3x	six	3rd cycle
3d	4d	decade	4th cycle
4p	4x	six	4th cycle
4s	4b	binary	4th cycle
4d	5d	decade	5th cycle
5p	5x	six	5th cycle
5s	5b	binary	5th cycle
4f	6f	fourteen	6th cycle
5d	6d	decade	6th cycle
6p	6x	six	6th cycle
6s	6b	binary	6th cycle
5f	7f	fourteen	7th cycle
6d	7d	decade	7th cycle
7p	7x	six	7th cycle
7s	7b	binary	7th cycle

5.10 New order for the naming convention of the electron rings

This list defines the transition in both rings and order.
This is the order of rings from the nucleus.

Old	New	type	cycle
1s	1b	binary	1st cycle
2p	2x	six	2nd cycle
2s	2b	binary	2nd cycle
3p	3x	six	3rd cycle
3s	3b	binary	3rd cycle
3d	4d	decade	4th cycle
4p	4x	six	4th cycle
4s	4b	binary	4th cycle
4d	5d	decade	5th cycle
5p	5x	six	5th cycle
5s	5b	binary	5th cycle
4f	6f	fourteen	6th cycle
5d	6d	decade	6th cycle
6p	6x	six	6th cycle
6s	6b	binary	6th cycle
5f	7f	fourteen	7th cycle
6d	7d	decade	7th cycle
7p	7x	six	7th cycle
7s	7b	binary	7th cycle

The 1st cycle is only a b ring (with up to 2 electrons).

The 2nd and 3rd cycles are bx rings (with 8 electrons).

The 4th and 5th cycles are bdx rings (with 18).

The 6th cycle and beyond are bfdx rings (with 32).

Element 118 completed the first 7 cycles or all of its rings through 7p or 7x.

The author expects the bfdx sequence in a cycle, with its 32 electrons, will continue repeating through subsequent elements which are not yet found in nature. The detection of elements heavier than 118 is unlikely. To investigate them, requires a better method using controlled fusion by extreme pressure, or using ransmutation, if that trick being used by nature is achieved in a laboratory. Creating heavy elements by high velocity collisions seems to get isotopes lacking stability. There are various combinations of protons and neutrons which are more stable than similar combinations. The mechanism being used must be closer to that found in the universe to explain the observed distribution of elements in the stars, planets, moons, and asteroids.

5.11 Simple illustration of the Practical Atomic Model

Wikipedia has an image for the electron configuration of sodium, element 11. This is from the topic Quantum Defect:

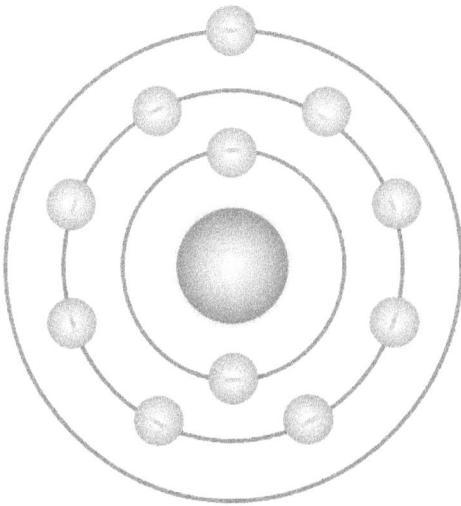

Its caption:
In an idealized Bohr model alkali atom (such as sodium, pictured here), the single outer-shell electron stays outside the ionic core and it would be expected to behave just as if in the same orbital of a hydrogen atom.

Observation:

This illustration matches the atom expected, simply because all orbitals are circular, and not having odd shapes.

The caption declares an unjustified assumption. The hydrogen atom has a single electron. The sodium atom has 10 electrons orbiting around the nucleus, inside this outer-shell electron. The radius of this outer-shell orbit is the result of those intervening electrons. The expectation could be for a similar orbit but it might not be the same. By one measurement, they are not the same. The hydrogen atom's electron orbit radius, when measured as a covalent radius is 28pm, but for sodium, the outer-shell orbit radius is 166 pm.

The words "behave just as if in the same orbital" are quite ambiguous.

6 Atomic Bonds

There are several bonds to describe.

First, the states of matter should be explained.

Excerpt from Wikipedia:

In physics, a state of matter is one of the distinct forms in which matter can exist. Four states of matter are observable in everyday life: solid, liquid, gas, and plasma. Many intermediate states are known to exist, such as liquid crystal, and some states only exist under extreme conditions, such as Bose–Einstein condensates, neutron-degenerate matter, and quark–gluon plasma, which only occur, respectively, in situations of extreme cold, extreme density, and extremely high energy. For a complete list of all exotic states of matter, see the list of states of matter. Historically, the distinction is made based on qualitative differences in properties. Matter in the solid state maintains a fixed volume and shape, with component particles (atoms, molecules or ions) close together and fixed into place. Matter in the liquid state maintains a fixed volume, but has a variable shape that adapts to fit its container. Its particles are still close together but move freely. Matter in the gaseous state has both variable volume and shape, adapting both to fit its container. Its particles are neither close together nor fixed in place. Matter in the plasma state has variable volume and shape, and contains neutral atoms as well as a significant number of ions and electrons, both of which can move around freely.

(Excerpt end)

Each state will be described and how it affects the atomic model. Plasma has unique behaviors and those are not relevant to this book's emphasis on the atomic model.

6.1 Gas

Excerpt from Wikipedia:

Gas is one of the four fundamental states of matter (the others being solid, liquid, and plasma). A pure gas may be made up of individual atoms (e.g. a noble gas like neon), elemental molecules made from one type of atom (e.g. oxygen), or compound molecules made from a variety of atoms (e.g. carbon dioxide). A gas mixture, such as air, contains a variety of pure gases. What distinguishes a gas from liquids and solids is the vast separation of the individual gas particles. This separation usually makes a colorless gas invisible to the human observer. The gaseous state of matter occurs between the liquid and plasma states, the latter of which provides the upper temperature boundary for gases.

(Excerpt end)

Observation:

At high temperatures, atoms and molecules can lose outer electrons, becoming positive ions. Any particle having a charge is considered plasma. Beyond Earth's ionosphere nearly everything is charged or is plasma. High energy radiation (UV and higher) will ionize exposed neutral matter.

Excerpt continued:

The only chemical elements that are
stable diatomic homonuclear molecules at STP are
hydrogen (H2), nitrogen (N2), oxygen (O2), and
two halogens: fluorine (F2) and chlorine (Cl2). When
grouped together with the monatomic noble gases –
 helium (He), neon (Ne), argon (Ar), krypton (Kr), xenon (Xe)
, and radon (Rn) – these gases are called "elemental
gases".

(Excerpt end),

Observation:

STP is standard temperature and pressure. Most
elements, other than the nobles, do not exist as a gas
without being bound to another atom, as a molecule of 2
or more atoms.

6.2 Liquid

Excerpt from Wikipedia:

A liquid is a nearly incompressible fluid that conforms to the shape of its container but retains a (nearly) constant volume independent of pressure. As such, it is the only state with a definite volume but no fixed shape. A liquid is made up of tiny vibrating particles of matter, such as atoms, held together by intermolecular bonds. Like a gas, a liquid is able to flow and take the shape of a container. Most liquids resist compression, although others can be compressed. Unlike a gas, a liquid does not disperse to fill every space of a container, and maintains a fairly constant density. A distinctive property of the liquid state is surface tension, leading to wetting phenomena. Water is, by far, the most common liquid on Earth.
The density of a liquid is usually close to that of a solid, and much higher than in a gas. Therefore, liquid and solid are both termed condensed matter. On the other hand, as liquids and gases share the ability to flow, they are both called fluids.

(Excerpt end)

Observation:

Solid follows for comparison. They have similar molecular bonds.

6.3 Solid

Excerpt from Wikipedia:

The molecules in a solid are closely packed together and contain the least amount of kinetic energy. A solid is characterized by structural rigidity and resistance to a force applied to the surface. Unlike a liquid, a solid object does not flow to take on the shape of its container, nor does it expand to fill the entire available volume like a gas. The atoms in a solid are bound to each other, either in a regular geometric lattice (crystalline solids, which include metals and ordinary ice), or irregularly (an amorphous solid such as common window glass). Solids cannot be compressed with little pressure.

(Excerpt end)

Observation:

The difference between a liquid and solid is the strength of the interaction between molecules.

6.4 Chemical Bonds

There are several relevant bonds here.

Excerpt from Wikipedia:

A chemical bond is a lasting attraction between atoms, ions or molecules that enables the formation of chemical compounds. The bond may result from the electrostatic force of attraction between oppositely charged ions as in ionic bonds or through the sharing of electrons as in covalent bonds. The strength of chemical bonds varies considerably; there are "strong bonds" or "primary bonds" such as covalent, ionic and metallic bonds, and "weak bonds" or "secondary bonds" such as dipole–dipole interactions, the London dispersion force and hydrogen bonding.
Since opposite charges attract via a simple electromagnetic force, the negatively charged electrons that are orbiting the nucleus and the positively charged protons in the nucleus attract each other. An electron positioned between two nuclei will be attracted to both of them, and the nuclei will be attracted toward electrons in this position. This attraction constitutes the chemical bond. Due to the matter wave nature of electrons and their smaller mass, they must occupy a much larger amount of volume compared with the nuclei, and this volume occupied by the electrons keeps the atomic nuclei in a bond relatively far apart, as compared with the size of the nuclei themselves.

(Excerpt end)

Observation:

Every mention of a electron requires it to be a real particle, which it is. There is absolutely no basis for a wave or probability behavior here.

An electron must be treated as a fundamental particle, not as an uncertain wave of probability moving in odd orbitals in a cloud.

6.5 Covalent Bond

This is the most relevant bond.

Excerpt from Wikipedia:

A covalent bond is a chemical bond that involves the sharing of electron pairs between atoms. These electron pairs are known as shared pairs or bonding pairs, and the stable balance of attractive and repulsive forces between atoms, when they share electrons, is known as covalent bonding. For many molecules, the sharing of electrons allows each atom to attain the equivalent of a full outer shell, corresponding to a stable electronic configuration. In organic chemistry, covalent bonds are much more common than ionic bonds.
Covalent bonding also includes many kinds of interactions, including σ-bonding, π-bonding, metal-to-metal bonding, agostic interactions, bent bonds, three-center two-electron bonds and three-center four-electron bonds.

(Excerpt end)

6.6 Chemical Kinetics

Chemical reactions require collisions of the respective electron clouds for the atoms to share electrons.

Excerpt from Wikipedia:

Chemical kinetics, also known as reaction kinetics, is the branch of physical chemistry that is concerned with understanding the rates of chemical reactions. It is to be contrasted with thermodynamics, which deals with the direction in which a process occurs but in itself tells nothing about its rate. Chemical kinetics includes investigations of how experimental conditions influence the speed of a chemical reaction and yield information about the reaction's mechanism and transition states, as well as the construction of mathematical models that also can describe the characteristics of a chemical reaction.

The physical state (solid, liquid, or gas) of a reactant is also an important factor of the rate of change. When reactants are in the same phase, as in aqueous solution, thermal motion brings them into contact. However, when they are in separate phases, the reaction is limited to the interface between the reactants. Reaction can occur only at their area of contact; in the case of a liquid and a gas, at the surface of the liquid. Vigorous shaking and stirring may be needed to bring the reaction to completion. This means that the more finely divided a solid or liquid reactant the greater its surface area per unit volume and the more contact it with the other reactant, thus the faster the reaction. To make an analogy, for example, when one starts a fire, one uses wood chips and small branches — one does not start with large logs right away. In organic chemistry, on water reactions are the exception to the rule that homogeneous reactions take place faster than heterogeneous reactions (are those reactions in which solute and solvent not mix properly)

The reactions are due to collisions of reactant species. The frequency with which the molecules or ions collide depends upon their concentrations. The more crowded the molecules are, the more likely they are to collide and react with one another. Thus, an increase in the concentrations of the reactants will usually result in the corresponding increase in the reaction rate, while a decrease in the concentrations will usually have a reverse effect. For example, combustion will occur more rapidly in pure oxygen than in air (21% oxygen).

Temperature usually has a major effect on the rate of a chemical reaction. Molecules at a higher temperature have more thermal energy. Although collision frequency is greater at higher temperatures, this alone contributes only a very small proportion to the increase in rate of reaction. Much more important is the fact that the proportion of reactant molecules with sufficient energy to react (energy greater than activation energy: $E > Ea$) is significantly higher and is explained in detail by the Maxwell–Boltzmann distribution of molecular energies.

At a given temperature, the chemical rate of a reaction depends on the value of the A-factor, the magnitude of the activation energy, and the concentrations of the reactants. Usually, rapid reactions require relatively small activation energies.
The 'rule of thumb' that the rate of chemical reactions doubles for every 10 °C temperature rise is a common misconception. This may have been generalized from the special case of biological systems, where
the α (temperature coefficient) is often between 1.5 and 2.5.

Increasing the pressure in a gaseous reaction will increase the number of collisions between reactants, increasing the rate of reaction. This is because the activity of a gas is directly proportional to the partial pressure of the gas. This is similar to the effect of increasing the concentration of a solution.

In addition to this straightforward mass-action effect, the rate coefficients themselves can change due to pressure. The rate coefficients and products of many high-temperature gas-phase reactions change if an inert gas is added to the mixture; variations on this effect are called fall-off and chemical activation. These phenomena are due to exothermic or endothermic reactions occurring faster than heat transfer, causing the reacting molecules to have non-thermal energy distributions (non-Boltzmann distribution). Increasing the pressure increases the heat transfer rate between the reacting molecules and the rest of the system, reducing this effect.

(Excerpt end)

Observation:

Perhaps, the excerpts are unnecessary. There is a rather simple conclusion:

Chemical reactions require the collisions of the respective atomic electron clouds.

Increasing the particle concentrations or increasing the velocity of the particles will increase the rate of molecular bonds. Increasing both will increase the rate further.

The respective rings of electrons in each atom must be in a compatible configuration for a bond to be achieved.

A noble gas element having a full outer ring will resist, by the electrostatic force, the intrusion of another electron cloud.

7 Radioactive Decay

An atom exists in a state of equilibrium.

An atom can make changes in an instant. These changes can be only in the electrons, only in the nucleus, or in both.

The equilibrium is disturbed and corrected within an instant.

Radioactive decay is one such behavior.

Among its instantaneous changes:

Alpha decay,
Beta decay,
Gamma decay,
Electron capture.

Each item will be described in turn.

Before the individual behaviors, there is another critical behavior to note. There is a particle competition within the nucleus. A context in the nucleus determines its decay selection.

Excerpt from Wikipedia:

Usually unstable nuclides are clearly either "neutron rich" or "proton rich", with the former undergoing beta decay and the latter undergoing electron capture (or more rarely, due to the higher energy requirements, positron decay). However, in a few cases of odd-proton, odd-neutron radionuclides, it may be energetically favorable for the radionuclide to decay to an even-proton, even-neutron isobar either by undergoing beta-positive or beta-negative decay.

An often-cited example is the single isotope ^{64}Cu (29 protons, 35 neutrons), which illustrates three types of beta decay in competition.

Copper-64 has a half-life of about 12.7 hours. This isotope has one unpaired proton and one unpaired neutron, so either the proton or the neutron can decay. This particular nuclide (though not all nuclides in this situation) is almost equally likely to decay through proton decay by positron emission (18%) or electron capture (43%) to ^{64}Ni, as it is through neutron decay by electron emission (39%) to ^{64}Zn.

(Excerpt end)

Observation:

Whether a nucleus is proton-rich or neutron-rich determines which beta decay occurs. Both are described below.

7.1 Electron capture

The electron capture step of radioactive decay involves a proton in the nucleus capturing one of the electrons in orbit at some distance.

The capture results in a) the proton changing into a neutron, b) a drop in the nucleus positive charge occurring at the same instant with c) the drop in the electron count, d) the difference between the number of electrons and protons remains the same.

The electron capture step can be important when anyone is using certain isotopes.

From Wikipedia: "The isotope technetium-97 decays only by electron capture, and could be inhibited from radioactive decay by fully ionizing it."

The electron capture event indicates there is a type of equilibrium between a nucleus and its set of electrons. There 2 sets of participants, with a number of protons matching the number of electrons, if the atom was neutral, not ionized, at the instant of the capture.

The trigger for any radioactive step is unknown, as well as the cause of an observed half-life, rather than some other duration. This ^{97}Tc nucleus can wait a long time before it pulls in an electron. ^{97}Tc has a half-life of 4.21×10^6 y.

7.2 Other radioactive behaviors

Several behaviors were identified above:

Alpha decay,
Beta decay,
Gamma decay,

7.2.1 Alpha Decay

Alpha decay requires the nucleus has the structure where an alpha particle already exists on the periphery of the nucleus. At the moment of instability, this particle of 2 protons bound to 2 neutrons is ejected by the Coulomb's force between a positive alpha particle and the rest of the nucleus which is also positive. The force for this ejection is sometimes called the weak force. Alpha decay occurs in the heaviest elements, apparently starting at Tellurium isotopes which have 59 protons. Several of its isotopes having 52 or more neutrons do alpha decay.

From the reference of radioactive elements, a general conclusion arises.

Among those lighter than Lead with alpha decay and a very long life before decay are 144Nd, 147Sm, 151Eu, 152Gd, 154Dy, 182Hf, 186Os, 190Pt.

Alpha decay more becomes frequent with heavy elements like Thorium or heavier.

7.2.2 Beta Decays

Beta decay requires the nucleus change its charge by one electron charge, by either an increase in net charge by emitting 1 e^- (beta minus) or a decrease by emitting 1 e^+ (beta plus).

Each will be explained further below. Sometimes, beta decay has a gamma ray.

7.2.3 Gamma Decay

Gamma decay is not clearly described. This step is usually associated with the element radium. It is also often associated with the alpha decay.

Excerpt from Wikipedia:

A sample of radium metal maintains itself at a higher temperature than its surroundings because of the radiation it emits – alpha particles, beta particles, and gamma rays.

More specifically, natural radium (which is mostly ^{226}Ra) emits mostly alpha particles, but other steps in its decay chain (the uranium or radium series) emit alpha or beta particles, and almost all particle emissions are accompanied by gamma rays.

In 2013, it was discovered that the nucleus of radium-224 is pear-shaped. This was the first discovery of an asymmetric nucleus.

(Excerpt end)

Observation:

The radium excerpt offers insight into radioactive decay.

First, a nucleus lacking symmetry enables a loss in stability when equilibrium between forces is disturbed.

Second, a spectrum is never provided for the gamma ray detection.

Synchrotron radiation when the motion of charged particles is diverted by a magnetic field. The peak frequency is determined by the velocity of the particles. This the mechanism for generating X-rays on Earth, like for medical imaging.

Alpha decay is a charged particle, having 2 protons and 2 neutrons, being ejected at high velocity, so this is charge in motion.

From Wikipedia about the alpha particle:

"Due to the mechanism of their production in standard alpha radioactive decay, alpha particles generally have a kinetic energy of about 5 MeV, and a velocity in the vicinity of 4% of the speed of light."

If the spectrum of radiation from radium were ever measured, it will be synchrotron radiation which is a relatively flat wave length distribution with infrared at the high end and by this observation the gamma ray wavelength is present in the mix at the low end. Radium is also described to radiate heat which is expected with infrared included in the distribution.

The alpha particles apparently have the great velocity required for the energy of gamma rays to be propagated by their ejection from the nucleus.

7.2.4 Beta Plus Decay

There are 2 beta decays, beta-plus or beta-minus.

Beta plus decay is also called positron emission.

Excerpt from Wikipedia:

Positron emission or beta plus decay (β+ decay) is a subtype of radioactive decay called beta decay, in which a proton inside a radionuclide nucleus is converted into a neutron while releasing a positron and an electron neutrino (v_e). Positron emission is mediated by the weak force. The positron is a type of beta particle (β+), the other beta particle being the electron (β−) emitted from the β− decay of a nucleus.

An example of positron emission (β+ decay) is magnesium-23 decaying into sodium-23.

Because positron emission decreases proton number relative to neutron number, positron decay happens typically in large "proton-rich" radionuclides. Positron decay results in nuclear transmutation, changing an atom of one chemical element into an atom of an element with an atomic number that is less by one unit.

A positron is ejected from the parent nucleus, and the daughter (Z−1) atom must shed an orbital electron to balance charge. The overall result is that the mass of two electrons is ejected from the atom (one for the positron and one for the electron).

(Excerpt end)

Observation:

This behavior is like electron capture.
Both are protons -1 and neutrons +1

The only apparent difference is the positron emission.

This description states a neutron is created while a positron is ejected and an electron is "shed" which suggests it is missing, so the final atom is an ion.

This is the scenario with ^{23}Mg and ^{23}Na

Start :

^{23}Mg has 12p + 11n, 12 e in orbit

End:
^{23}Na has 11p + 12n, 10 e in orbit because 1 electron was "shed"
Positron was ejected also

The instantaneous event when everything is simultaneous:
a) 1e was captured from K shell by a proton to become a neutron.
b) This instant has the 12p in nucleus +1 compared to remaining 11 electrons in orbit.

c) This captured electron's polarity is flipped from normal e^- to e^+ becoming a positron.

d) Ejecting that "flipped in an instant" positron at +1 balances the charges in the atom.

e) Therefore, the positron is the captured electron but its polarity was flipped. The positron did not come from a proton as currently described.

f) Another, second electron from K shell is captured.

g) This instant has the nucleus at 11p + 12p again

The stated changes are from 12 to 10 electrons in orbit, so one electron is captured to form a neutron, while one positron is ejected for mass balance.

At the moment an electron flips its polarity while adjacent to a proton, the two + charges repel, causing the positron ejection.

^{23}Na is known to decay with β^+ emission having a half-life of 22.422 seconds.

Positron creation is also part of Particle Pair Production, which is covered in section 10 Light.

Therefore, based on its combination of actions, beta plus decay can be called a double electron capture but the second is ejected from the nucleus as a positron.

7.2.5 Beta-minus decay

As noted above with nucleus competition description, beta minus decay occurs in a "neutron rich" nucleus. As a result, one neutron ejects its accompanying electron due to too many electrons among the neutrons. Equilibrium among these electrons in the nucleus is restored by ejecting one of the electrons.

Excerpt from Wikipedia:

> In nuclear physics, beta decay (β-decay) is a type of radioactive decay in which a beta particle (fast energetic electron or positron) is emitted from an atomic nucleus, transforming the original nuclide to an isobar. For example, beta decay of a neutron transforms it into a proton by the emission of an electron accompanied by an antineutrino.

(Excerpt end)

Observation:

Beta minus decay is accompanied by an antineutrino. The neutrino was described in PPPB as a particle having a questionable existence. It has no mass to use for measurement and confirmation. The only method used for its detection is a deuterium atom (in deep pools of heavy water) losing its neutron. It is impossible to use that observation as evidence when there could be other causes for that separation, as yet unidentified.

An indirect method of detection requires absolute certainty of no other causes than the one being pursued. LIGO is the notorious example of a mistake with indirect detection. LIGO neglected the predictable, terrestrial source for a crust disturbance. That mistake resulted in never confirmed claims of a distant astrophysical source generating non-existent gravitational waves. An indirect method requires thoroughness to avoid a false conclusion.

7.3 Even and odd Nucleus

The section 4 Build Nucleus described how the even or odd count of nucleons directly affects the number of electrons required for a stable nucleus.

Radioactive decay is also driven by this even or odd simply because decay is the opposite of stability. If the electron count in the nucleus is not correct for stability then the nucleus reacts in an attempt to achieve stability. The author created a data set containing all of the stable isotopes and many of the radioactive isotopes. This file is identified in section 3 Data Sets. There are 118 elements so this data set includes over 1000 isotopes. Instead of this book detailing every radioactive isotope and the details of its action, only a few are described here. A reader wanting more is free to reference the data set having all.

In these descriptions p and e refer to protons and electrons in the nucleus. Z is commonly used for the atomic number (the number of protons with no attached electron) and N for neutrons (which is an attached pair of proton and electron). Behaviors in the nucleus are clearer when not mentioning neutrons. A neutron is not a fundamental particle like its 2 components.

The first example has an odd atomic number.

The second example has an even atomic number.

7.3.1 Example 1 Potassium

Potassium has an odd atomic number, 19.
Potassium has 8 isotopes noted in the data set. Other isotopes of potassium would have very brief half lives and are not important to the conclusion.

A first impression of the 8 notes there are only 3 isotopes which are stable or have a long half-life.

^{37}K has 37p +18e and decays by B+ (which is a capture of 2 electrons by the nucleus and the ejection of 1 positron, or a B+ particle from the nucleus) in 24s. Its remnant is ^{37}Ar which is unstable.

^{38}K has 38p +19e and decays by B+ in 8m. Its remnant is ^{38}Ar which is stable.

^{39}K has 39p+20e and is stable.

^{40}K has 40p +21e and decays by B- (which is an electron ejected from the nucleus) in 3×10^9y. Its remnant is ^{38}Ar which is stable.

^{41}K has 41p+22e and is stable.

^{42}K has 42p +23e and decays by B- (which is an electron ejected from the nucleus) in 12h. Its remnant is ^{42}Ca which is stable.

^{43}K has 43p +24e and decays by B- (which is an electron ejected from the nucleus) in 22h. Its remnant is ^{43}Ca which is stable.

^{44}K has 44p +25e and decays by B- (which is an electron ejected from the nucleus) in 22m. Its remnant is ^{44}Ca which is stable.

Observations:

^{39}K is what the author will call the middle stable isotope. Isotopes having a lower atomic number than this middle stable one are known as proton-rich nuclei so they react by reducing the positive charge in the nucleus. B+ decay reduces Z by 1. This results in the next lower atomic number.

Potassium is the element having an odd Z at 19. Its isotopes have different counts of neutrons for their atomic weight (Z + N). A nucleon count is the same as the atomic weight.

Odd nucleon counts of 37 to 43 have an even count of electrons resulting in an odd Z, 19 for potassium. Of the 4 odds, only ^{39}K and ^{41}K are stable; ^{37}K and ^{43}K are unstable Even nucleon counts of 38, 40, 42 have an odd count of electrons resulting in an odd Z, 19 for potassium. All 3 evens are unstable.

7.3.2 Example 2 Calcium

Calcium has an even atomic number, 20.
Calcium has 11 isotopes noted in the data set. Other isotopes of calcium would have very brief half lives and are not important to the conclusion.

A first impression of the 11 notes there are 8 isotopes which are stable or have a long half-life. Nuclei with an even count of nucleons tend to be more stable than those with an odd count.

^{39}Ca has 39p +19e and decays by B+ in 0.86s. Its remnant is ^{39}K which is stable.

^{40}Ca has 40p+20e and is stable.

^{41}Ca has 41p+21e and decays by electron capture in 10y. Its remnant is ^{41}K which is stable

^{42}Ca has 42p+22e and is stable.

^{43}Ca has 43p+23e and is stable.

^{44}Ca has 44p+24e and is stable.

^{45}Ca has 45p+25e and decays by B- (or electron ejection) in 163d. Its remnant is ^{45}Sc which is stable.

^{46}Ca has 46p+26e and is stable.

^{47}Ca has 47p+27e and decays by B- (or electron ejection) in 4.7d. Its remnant is 47Sc which is unstable.

^{48}Ca has 48p+28e and decays by 2x B- (or 2 electrons are ejected) in 6×10^{19}y. Its remnant is 48Sc which is unstable.

^{49}Ca has 49p+29e and decays by B- (or electron ejection) in 8.7m. Its remnant is 49Sc which is unstable.

Observations:

^{43}Ca is what the author will call the middle stable isotope. Radioactive isotopes having a lower atomic number than this middle stable one are known as proton-rich nuclei so they react by reducing the positive charge in the nucleus. B+ decay reduces Z by 1. This results in the next lower atomic number.
Radioactive isotopes having a higher atomic number than this middle stable one are known as neutron-rich nuclei so they react by decreasing the net charge in the nucleus by the B- decay which increases Z by 1. This results in the next higher atomic number. If this ejected electron is captured by the cloud then the atom's net charge is maintained; or a) the neutral atom became a positive ion; or b) the ion increased its net + charge by one.

Calcium is the element having an even Z at 20. Its isotopes have different counts of neutrons for their atomic weight (Z + N). A nucleon count is the same as the atomic weight.

Even nucleon counts of 40, 42, 44, 46, 48 have an even count of electrons resulting in an even Z, 20 for calcium. Of these 5 even isotopes, the first 4 are stable, while ^{48}Ca decays by ejecting 2 electrons after many years.

Odd nucleon counts of 39, 41, 43, 45, 47, 49 have an odd count of electrons resulting in an even Z, 20 for calcium.

Of these 6 odd isotopes, only ^{43}Ca is stable while the other 5 odd isotopes decay.

There are several radioactive isotopes among the 118 elements which exhibit this behavior: With a long half-life, the action taken is more than might be expected. Both ^{48}Ca and ^{49}Ca exhibit this behavior because they execute the expected decay twice, not once, and both have half-lives spanning many years.
For some elements taking the alpha particle ejection step, sometimes they have very long half-lives.

One might note that in Wikipedia, both ^{40}Ca and ^{46}Ca are classified as Observationally Sable. This phrase is because there is another isotope (^{41}Ca and ^{45}Ca) which decays between this particular isotope and the middle stable one.

From Wikipedia: "Isotopes that are theoretically believed to be unstable but have not been observed to decay are termed as observationally stable. "

7.3.3 Summary of gathering

The even or odd number of nucleons in a nucleus affects its radioactive decay behaviors just like odd or even affects its stability, as noted in Section 4 Build Nucleus.

8 Periodic Table

Several behaviors for each element are described:

a) Stable isotopes,

b) Electron shells, using the new ring names, in their order from the nucleus.

c) Number of valence electrons by "VE="

d) Any anomalies.

The numbered sections in this Periodic Table section are outlined as 8.X where X is the atomic number. All 118 elements have their section.

8.1 Element 1 is Hydrogen or H

Hydrogen is the combination of 1 proton and 1 electron.

These are its isotopes which are stable or have a half-life over 1 year.

1H, 2H,
3H = 12.323y

This is its electron configuration using the new ring names.

1,H,1b1,VE=1

The hydrogen atom has been thoroughly researched.

Tthe single orbiting electron takes defined or quantized orbits based on the atom's energy level. Those studies lead to the Lyman and Balmer series as well as other series infrequently observed.

Astronomy and cosmology have difficulty accepting plasma physics involving protons and electrons not bound within atoms. Loose protons are often ionized hydrogen. A cloud of loose protons is called a H II region, where H I is a reference to neutral hydrogen (this state is rare in cosmology).

8.2 Element 2 is Helium or He

Helium has 2 protons and 2 neutrons.

These are its isotopes which are stable or have a half-life over 1 year.

3He, 4He

This is its electron configuration using the new ring names.

2,He,1b2,VE=2.

Helium is inert with only an opposing pair of electrons in a tight orbit having a measured covalent radius of only 31 pm.

8.3 Element 3 is Lithium or Li

Lithium has 3 protons and 4 neutrons.

These are its isotopes which are stable or have a half-life over 1 year.

6Li, 7Li

This is its electron configuration using the new ring names.

3,Li,1b2,2b1,VE=1.

8.4 Element 4 is Beryllium or Be

.

Beryllium has 4 protons and 5 neutrons.

These are its isotopes which are stable or have a half-life over 1 year.

9Be
10Be = 5E6 y

This is its electron configuration using the new ring names.

4,Be,1b2,2b2,VE=2.

8.5 Element 5 is Boron or B

Boron has 5 protons and 6 neutrons.

These are its isotopes which are stable or have a half-life over 1 year.

10B, 11B

This is its electron configuration using the new ring names.

5,B,1b2,2x1,2b2,VE=3.

8.6 Element 6 is Carbon or C

Carbon has 6 protons and 6 neutrons.

These are its isotopes which are stable or have a half-life over 1 year.

12C, 13C
14C = 7530 y

This is its electron configuration using the new ring names.

6 6,C,1b2,2x2,2b2,VE=4.

8.7 Element 7 is Nitrogen or N

Nitrogen has 7 protons and 7 neutrons.

These are its isotopes which are stable or have a half-life over 1 year.

14N, 15N

This is its electron configuration using the new ring names.

7,N,1b2,2x3,2b2,VE=5.

8.8 Element 8 is Oxygen or O

Oxygen has 8 protons and 8 neutrons.

These are its isotopes which are stable or have a half-life over 1 year.

16O, 17O, 18O

This is its electron configuration using the new ring names.

8,O,1b2,2x4,2b2,VE=6.

8.9 Element 9 is Fluorine or F

Fluorine has 9 protons and 10 neutrons.

These are its isotopes which are stable or have a half-life over 1 year.

19F

This is its electron configuration using the new ring names.

9, 9,F,1b2,2x5,2b2,VE=7.

8.10 Element 10 is Neon or Ne

Neon has 10 protons and 10 neutrons.

These are its isotopes which are stable or have a half-life over 1 year.

20Ne, 21Ne, 22Ne

This is its electron configuration using the new ring names.

These are its isotopes which are stable or have a half-life over 1 year.

This is its electron configuration using the new ring names.

10, 10,Ne,1b2,2x6,2b2,VE=8.

8.11 Element 11 is Sodium or Na

Sodium has 11 protons and 12 neutrons.

These are its isotopes which are stable or have a half-life over 1 year.

23Na

This is its electron configuration using the new ring names.
11,Na,1b2,2x6,2b2,3b1,VE=1.

8.12 Element 12 is Magnesium or Mg

Magnesium has 12 protons and 12 neutrons.

These are its isotopes which are stable or have a half-life over 1 year.

24Mg, 25Mg, 26Mg

This is its electron configuration using the new ring names.

12,Mg,1b2,2x6,2b2,3b2,VE=2.

8.13 Element 13 is Aluminum or Al

Aluminum has 13 protons and 14 neutrons.

These are its isotopes which are stable or have a half-life over 1 year.

27Al

This is its electron configuration using the new ring names.

13,Al,1b2,2x6,2b2,3x1,3b2,VE=3.

8.14 Element 14 is Silicon or Si

Silicon has 14 protons and 14 neutrons.

These are its isotopes which are stable or have a half-life over 1 year.

28Si, 29Si, 30Si

This is its electron configuration using the new ring names.

14,Si,1b2,2x6,2b2,3x2,3b2,VE=4.

8.15 Element 15 is Phosphorus or P

Phosphorus has 15 protons and 16 neutrons.

These are its isotopes which are stable or have a half-life over 1 year.

31P

This is its electron configuration using the new ring names.

15,P,1b2,2x6,2b2,3x3,3b2,VE=5.

8.16 Element 16 is Sulfur or S

Sulfur has 16 protons and 16 neutrons.

32S, 33S, 34S

These are its isotopes which are stable or have a half-life over 1 year.

This is its electron configuration using the new ring names.

16,S,1b2,2x6,2b2,3x4,3b2,VE=6.

8.17 Element 17 is Chlorine or Cl

Chlorine has 17 protons and 18 neutrons.

These are its isotopes which are stable or have a half-life over 1 year.

35Cl
36Cl = 3E5 y

This is its electron configuration using the new ring names.

17,Cl,1b2,2x6,2b2,3x5,3b2,VE=7.

8.18 Element 18 is Argon or Ar

Argon has 18 protons and 22 neutrons.

These are its isotopes which are stable or have a half-life over 1 year.

40Ar
42Ar = 33 y

This is its electron configuration using the new ring names.

18,Ar,1b2,2x6,2b2,3x6,3b2,VE=8.

8.19 Element 19 is Potassium or K

Potassium has 19 protons and 20 neutrons.

These are its isotopes which are stable or have a half-life over 1 year.

39K

This is its electron configuration using the new ring names.

19,K,1b2,2x6,2b2,3x6,3b2,4b1,VE=1.

8.20 Element 20 is Calcium or Ca

Calcium has 20 protons and 20 neutrons.

These are its isotopes which are stable or have a half-life over 1 year.

40Ca, 42Ca, 43Ca, 44Ca, 46Ca
41Ca = 10 y
Ca48 = 6E9 y

This is its electron configuration using the new ring names.

20,Ca,1b2,2x6,2b2,3x6,3b2,4b2,VE=2.

8.21 Element 21 is Scandium or Sc

Scandium has 21 protons and 24 neutrons.

These are its isotopes which are stable or have a half-life over 1 year.

45Sc

This is its electron configuration using the new ring names.

21,Sc,1b2,2x6,2b2,3x6,3b2,4d1,4b2,VE=2.

8.22 Element 22 is Titanium or Ti

Titanium has 22 protons and 26 neutrons.

46Ti, 47Ti, 48Ti, 49Ti, 50Ti

These are its isotopes which are stable or have a half-life over 1 year.

This is its electron configuration using the new ring names.

22,Ti,1b2,2x6,2b2,3x6,3b2,4d2,4b2,VE=2.

8.23 Element 23 is Vanadium or V

Vanadium has 23 protons and 28 neutrons.

These are its isotopes which are stable or have a half-life over 1 year.
51V
50V = 1.4E17 y

This is its electron configuration using the new ring names.

23,V,1b2,2x6,2b2,3x6,3b2,4d3,4b2,VE=2.

8.24 Element 24 is Chromium or Cr

Chromium has 24 protons and 28 neutrons.

These are its isotopes which are stable or have a half-life over 1 year.

50Cr, 52Cr, 53Cr, 54Cr

This is its electron configuration using the new ring names.

24,Cr,1b2,2x6,2b2,3x6,3b2,4d5,4b1,VE=6.

This configuration has an anomaly from expected, with an electron moved between adjacent rings.

The change from expected: 4d4, 4b2 is 4d5,4b1.

8.25 Element 25 is Manganese or Mn

Manganese has 25 protons and 30 neutrons.

These are its isotopes which are stable or have a half-life over 1 year.

55Mn

This is its electron configuration using the new ring names.

25,Mn,1b2,2x6,2b2,3x6,3b2,4d5,4b2,VE=2.

8.26 Element 26 is Iron or Fe

Iron has 26 protons and 30 neutrons.

These are its isotopes which are stable or have a half-life over 1 year.

54Fe, 56Fe, 57Fe, 58Fe
55Fe = 2.7 y
60Fe = 5E6 y

This is its electron configuration using the new ring names.

26,Fe,1b2,2x6,2b2,3x6,3b2,4d6,4b2,VE=2.

8.27 Element 27 is Cobalt or Co

Cobalt has 27 protons and 32 neutrons.

These are its isotopes which are stable or have a half-life over 1 year.

59Co
60Co = 271 y

This is its electron configuration using the new ring names.

27,Co,1b2,2x6,2b2,3x6,3b2,4d7,4b2,VE=2.

8.28 Element 28 is Nickel or Ni
Nickel has 28 protons and 31 neutrons.

These are its isotopes which are stable or have a half-life over 1 year.

60Ni, 61Ni, 62Ni, 64Ni
63Ni = 8E4 y

This is its electron configuration using the new ring names.

28,Ni,1b2,2x6,2b2,3x6,3b2,4d8,4b2,VE=2.

8.29 Element 29 is Copper or Cu

Copper has 29 protons and 35 neutrons.

These are its isotopes which are stable or have a half-life over 1 year.

63Cu, 65Cu
(64Cu = 12 h)

This is its electron configuration using the new ring names.
29,Cu,1b2,2x6,2b2,3x6,3b2,4d10,4b1,VE=6.

This configuration has an anomaly from expected, with an electron moved between adjacent rings.

The change from expected: 4d9, 4b2 is 4d10,4b1.

8.30 Element 30 is Zinc or Zn

Zinc has 30 protons and 35 neutrons.

These are its isotopes which are stable or have a half-life over 1 year.

64Zn, 65Zn, 66Zn, 67Zn, 68Zn, 70Zn

This is its electron configuration using the new ring names.

330,Zn,1b2,2x6,2b2,3x6,3b2,4d10,4b2,VE=2.

8.31 Element 31 is Gallium or Ga

Gallium has 31 protons and 39 neutrons.
These are its isotopes which are stable or have a half-life over 1 year.

69Ga, 71Ga
(70Ga = 11.4d)

This is its electron configuration using the new ring names.

1,Ga,1b2,2x6,2b2,3x6,3b2,4d10,4x1,4b2,VE=3.

8.32 Element 32 is Germanium or Ge

Germanium has 32 protons and 41 neutrons.

These are its isotopes which are stable or have a half-life over 1 year.

70Ge, 72Ge, 73Ge, 74Ge
76Ge = 2E4 y

This is its electron configuration using the new ring names.

2,Ge,1b2,2x6,2b2,3x6,3b2,4d10,4x2,4b2,VE=4.

8.33 Element 33 is Arsenic or As

Arsenic has 33 protons and 42 neutrons.

These are its isotopes which are stable or have a half-life over 1 year.

75As

This is its electron configuration using the new ring names.

33,As,1b2,2x6,2b2,3x6,3b2,4d10,4x3,4b2,VE=5.

8.34 Element 34 is Selenium or Se

Selenium has 34 protons and 45 neutrons.

74Se, 76Se, 77Se, 78Se, 80Se
79Se = 3E5 y
82Se = 120 y

These are its isotopes which are stable or have a half-life over 1 year.

This is its electron configuration using the new ring names.

34,Se,1b2,2x6,2b2,3x6,3b2,4d10,4x4,4b2,VE=6.

8.35 Element 35 is Bromine or Br

Bromine has 35 protons and 45 neutrons.

These are its isotopes which are stable or have a half-life over 1 year.

79Br, 81Br
(80Br = 18 m)

This is its electron configuration using the new ring names.

35,Br,1b2,2x6,2b2,3x6,3b2,4d10,4x5,4b2,VE=7.

8.36 Element 36 is Krypton or Kr

Krypton has 36 protons and 48 neutrons.

These are its isotopes which are stable or have a half-life over 1 year.

82Kr, 83Kr, 84Kr, 86Kr
81Kr = 2E5 y
85Kr = 10.8 y

This is its electron configuration using the new ring names.

36,Kr,1b2,2x6,2b2,3x6,3b2,4d10,4x6,4b2,VE=8.

8.37 Element 37 is Rubidium ot Rb

Rubidium has 37 protons and 48 neutrons.

These are its isotopes which are stable or have a half-life over 1 year.

85Rb
87Rb = 5E10 y

37,Rb,1b2,2x6,2b2,3x6,3b2,4d10,4x6,4b2,5b1,VE=1.

8.38 Element 38 is Strontium or Sr

Strontium has 38 protons and 50 neutrons.

These are its isotopes which are stable or have a half-life over 1 year.

84Sr, 86Sr, 87Sr, 88Sr
90Sr = 29 y

This is its electron configuration using the new ring names.

38,Sr,1b2,2x6,2b2,3x6,3b2,4d10,4x6,4b2,5b2,VE=2.

8.39 Element 39 is Yttrium or Y

Yttrium has 39 protons and 50 neutrons.

These are its isotopes which are stable or have a half-life over 1 year.

89Y

This is its electron configuration using the new ring names.

39,Y,1b2,2x6,2b2,3x6,3b2,4d10,4x6,4b2,5d1,5b2,VE=2.

8.40 Element 40 is Zirconium or Zr

Zirconium has 40 protons and 51 neutrons.

These are its isotopes which are stable or have a half-life over 1 year.

90Zr, 91Zr, 92Zr, 94Zr
93Zr = 1.5E18 y
96Zr = 20E18 y

This is its electron configuration using the new ring names.

40,Zr,1b2,2x6,2b2,3x6,3b2,4d10,4x6,4b2,5d2,5b2,VE=2.

8.41 Element 41 is Niobium or Nb

Niobium has 41 protons and 52 neutrons.

These are its isotopes which are stable or have a half-life over 1 year.

93Nb
91Nb = 680 y
92Nb = 3.5E7 y
94Nb = 2E4 y

This is its electron configuration using the new ring names.

41,Nb,1b2,2x6,2b2,3x6,3b2,4d10,4x6,4b2,5d4,5b1,VE=5.

This configuration has an anomaly from expected, with an electron moved between adjacent rings.

The change from expected: 5d3, 5b2 is 5d4,5b1.

8.42 Element 42 is Molybdenum or Mo

Molybdenum has 42 protons and 54 neutrons.

These are its isotopes which are stable or have a half-life over 1 year.

92Mo, 94Mo, 95Mo, 96Mo, 97Mo, 98Mo
93Mo = 4 y
100Mo = 9E18 y

This is its electron configuration using the new ring names.

42,Mo,1b2,2x6,2b2,3x6,3b2,4d10,4x6,4b2,5d5,5b1,VE=6.

This configuration has an anomaly from expected, with an electron moved between adjacent rings.

The change from expected: 5d4, 5b2 is 5d5,5b1.

8.43 Element 43 is Technetium or Tc

Technetium has 43 protons and 55 neutrons.

These are its isotopes which are stable or have a half-life over 1 year.

97Tc = 4E6 y
98Tc = 4E6 y
99Tc = 2E5 y

This is its electron configuration using the new ring names.

43,Tc,1b2,2x6,2b2,3x6,3b2,4d10,4x6,4b2,5d5,5b2,VE=7.

8.44 Element 44 is Ruthenium or Ru

Ruthenium has 44 protons and 57 neutrons.

These are its isotopes which are stable or have a half-life over 1 year.

96Ru, 98Ru, 98Ru, 98Ru, 99Ru, 100Ru, 101Ru, 102Ru, 104Ru

This is its electron configuration using the new ring names.

44,Ru,1b2,2x6,2b2,3x6,3b2,4d10,4x6,4b2,5d7,5b1,VE=8.

This configuration has an anomaly from expected, with an electron moved between adjacent rings.

The change from expected: 5d6, 5b2 is 5d7,5b1.

8.45 Element 45 is Rhodium or Rh

Rhodium has 45 protons and 58 neutrons.

These are its isotopes which are stable or have a half-life over 1 year.

103Rh
101Rh = 3.3 y

This is its electron configuration using the new ring names.

45,Rh,1b2,2x6,2b2,3x6,3b2,4d10,4x6,5d8,5b1,VE=9

This configuration has an anomaly from expected, with an electron moved between adjacent rings.

The change from expected: 5d7, 5b2 is 5d8,5b1.

8.46 Element 46 is Palladium or Pd

Palladium has 46 protons and 61 neutrons.

These are its isotopes which are stable or have a half-life over 1 year.

102Pd, 104Pd, 105Pd, 106Pd, 108Pd, 110Pd
107Pd = 6.5E6 y

This is its electron configuration using the new ring names.

46,Pd,1b2,2x6,2b2,3x6,3b2,4d10,4x6,5d9,5b1, VE=10

This configuration has an anomaly from expected, with an electron moved between adjacent rings.

The change from expected: 5d8, 5b2 is 5d9,5b1.

8.47 Element 47 is Silver or Ag

Silver has 47 protons and 61 neutrons.

These are its isotopes which are stable or have a half-life over 1 year.

107Ag, 109Ag
(108Ag = 2.4 m)

This is its electron configuration using the new ring names.

47,Ag,1b2,2x6,2b2,3x6,3b2,4d10,4x6,4b2,5d10,5b1, VE=11

This configuration has an anomaly from expected, with an electron moved between adjacent rings.

The change from expected: 5d9, 5b2 is 5d10,5b1.

8.48 Element 48 is Cadmium or Cd

Cadmium has 48 protons and 64 neutrons.

These are its isotopes which are stable or have a half-life over 1 year.

106Cd, 108Cd, 110Cd, 111Cd, 112Cd, 114Cd
109Cd = 461 y
113Cd = 8E15 y
116Cd = 8E19 y

This is its electron configuration using the new ring names.

48,Cd,1b2,2x6,2b2,3x6,3b2,4d10,4x6,4b2,5d10,5b2, VE=2.

8.49 Element 49 is Indium or In

Indium has 49 protons and 66 neutrons.

These are its isotopes which are stable or have a half-life over 1 year.

113In
115In = 4.4E14 y

This is its electron configuration using the new ring names.

49,In,1b2,2x6,2b2,3x6,3b2,4d10,4x6,4b2,5d10,5x1,5b2, VE=3.

8.50 Element 50 is Tin or Sn

Tin has 50 protons and 69 neutrons.

These are its isotopes which are stable or have a half-life over 1 year.

112Sn, 114Sn, 115Sn, 116Sn, 117Sn, 118Sn, 119Sn, 120Sn, 122Sn, 124Sn

126Sn = 2.3E5 y

This is its electron configuration using the new ring names.

50,Sn,1b2,2x6,2b2,3x6,3b2,4d10,4x6,4b2,5d10,5x2,5b2, VE=4.

8.51 Element 51 is Antimony or Sb

Antimony has 51 protons and 71 neutrons.

These are its isotopes which are stable or have a half-life over 1 year.

121Sb, 123Sb

125Sb = 2.8 y

This is its electron configuration using the new ring names.

51,Sb,1b2,2x6,2b2,3x6,3b2,4d10,4x6,4b2,5d10,5x3,5b2, VE=5.

8.52 Element 52 is Tellurium or Te

Tellurium has 52 protons and 76 neutrons.

These are its isotopes which are stable or have a half-life over 1 year.

120Te, 122Te, 123Te, 124Te, 125Te, 126Te

128Te = 2.2E24 y
130Te = 8E20 y

This is its electron configuration using the new ring names.

52,Te,1b2,2x6,2b2,3x6,3b2,4d10,4x6,4b2,5d10,5x4,5b2,
VE=6.

8.53 Element 53 is Iodine or I

Iodine has 53 protons and 74 neutrons.

These are its isotopes which are stable or have a half-life over 1 year.

127I

129I = 1.6E7 y

This is its electron configuration using the new ring names.

53,I,1b2,2x6,2b2,3x6,3b2,4d10,4x6,4b2,5d10,5x5,5b2, VE=7

8.54 Element 54 is Xenon or Xe

Xenon has 54 protons and 77 neutrons.

These are its isotopes which are stable or have a half-life over 1 year.

126Xe, 128Xe, 129Xe, 130Xe, 131Xe, 132Xe, 134Xe

124Xe = 1.8E22 y
136Xe = 2.17E21 y

This is its electron configuration using the new ring names.

54,Xe,1b2,2x6,2b2,3x6,3b2,4d10,4x6,4b2,5d10,5x6,5b2,
VE=8.

8.55 Element 55 is Caesium or Cs

Cesium has 55 protons and 78 neutrons.

These are its isotopes which are stable or have a half-life over 1 year.

133Cs

135Cs = 2.3E6 y

This is its electron configuration using the new ring names.

55,Cs,1b2,2x6,2b2,3x6,3b2,4d10,4x6,4b2,5d10,5x6,5b2, 6b1,VE=1.

8.56 Element 56 is Barium or Ba

Barium has 56 protons and 81 neutrons.

These are its isotopes which are stable or have a half-life over 1 year.

132Ba, 134Ba, 135Ba, 136Ba, 137Ba, 138Ba

130Ba = 1.6E21 y
133Ba = 10.5 y

This is its electron configuration using the new ring names.

56,Ba,1b2,2x6,2b2,3x6,3b2,4d10,4x6,4b2,5d10,5x6,5b2, 6b2, VE=2.

8.57 Element 57 is Lanthanum or La

Lanthanum has 57 protons and 82 neutrons.

These are its isotopes which are stable or have a half-life over 1 year.

139La

137La = 6E4 y
138La = 1E11 y

This is its electron configuration using the new ring names.

57,La,1b2,2x6,2b2,3x6,3b2,4d10,4x6,4b2,5d10,5x6,5b2,
6d1,6b2,VE=2.

8.58 Element 58 is Cerium or Ce

Cerium has 58 protons and 82 neutrons.

These are its isotopes which are stable or have a half-life over 1 year.

136Ce, 138Ce, 140Ce, 142Ce

This is its electron configuration using the new ring names.

58,Ce,1b2,2x6,2b2,3x6,3b2,4d10,4x6,4b2,5d10,5x6,5b2, 6f2,6b2,VE=2.

8.59 Element 59 is Praseodymium or Pr

Praseodymium has 59 protons and 82 neutrons.

These are its isotopes which are stable or have a half-life over 1 year.

141Pr

This is its electron configuration using the new ring names.

59,Pr,1b2,2x6,2b2,3x6,3b2,4d10,4x6,4b2,5d10,5x6,5b2,
6f3,6b2,VE=2.

8.60 Element 60 is Neodymium or Nd

Neodymium has 60 protons and 84 neutrons.

These are its isotopes which are stable or have a half-life over 1 year.

142 Nd, 143Nd, 145Nd, 146Nd, 148Nd

144Nd = 2.3E15 y
150Nd = 6.7E18 y

This is its electron configuration using the new ring names.

60,Nd,1b2,2x6,2b2,3x6,3b2,4d10,4x6,4b2,5d10,5x6,5b2,
6f4,6b2,VE=2.

8.61 Element 61 is Promethium or Pm

Promethium has 61 protons and 84 neutrons.

These are its isotopes which are stable or have a half-life over 1 year.

145Pm = 17.7 y
146Pm = 5.5 y
147Pm = 2.6 y

This is its electron configuration using the new ring names.

61,Pm,1b2,2x6,2b2,3x6,3b2,4d10,4x6,4b2,5d10,5x6,
5b2,6f5,6b2,VE=2.

8.62 Element 62 is Samarium or Sm

Samarium has 62 protons and 88 neutrons.

These are its isotopes which are stable or have a half-life over 1 year.

144Sm, 149Sm, 150Sm, 152Sm, 154Sm

146Sm = 6.8E9 y
147Sm = 1.1E11 y
148Sm = 7E11 y
151Sm = 88.8 y

This is its electron configuration using the new ring names.

62,Sm,1b2,2x6,2b2,3x6,3b2,4d10,4x6,4b2,5d10,5x6,
5b2,6f6,6b2,VE=2.

8.63 Element 63 is Europium or Eu

Europium has 63 protons and 89 neutrons.

These are its isotopes which are stable or have a half-life over 1 year.

153Eu

150Eu = 36.9 y
151Eu = 4.E18 y
152Eu = 13.5 y
154Eu = 8.6 y
155Eu = 4.8 y

This is its electron configuration using the new ring names.

63,Eu,1b2,2x6,2b2,3x6,3b2,4d10,4x6,4b2,5d10,5x6,5b2,
6f7,6b2,VE=2.

8.64 Element 64 is Gadolinium or Gd

Gadolinium has 64 protons and 93 neutrons.

These are its isotopes which are stable or have a half-life over 1 year.

154Gd, 155Gd, 156Gd, 157Gd, 158Gd, 160Gd

148Gd = 4.8 y
150Gd = 1.8E6 y
152Gd = 1.1E4 y

This is its electron configuration using the new ring names.

64,Gd,1b2,2x6,2b2,3x6,3b2,4d10,4x6,4b2,5d10,5x6,
5b2,6f7,6d1,6b2,VE=2.

This configuration has an anomaly from expected, with an electron moved between adjacent rings.

The change from expected: 6f8, 6d0 is 6f7,6d1.

8.65 Element 65 is Terbium or Tb

Terbium has 65 protons and 94 neutrons.

These are its isotopes which are stable or have a half-life over 1 year.

159Tb

157Tb = 71 y
158Tb = 180 y

This is its electron configuration using the new ring names.

65,Tb,1b2,2x6,2b2,3x6,3b2,4d10,4x6,4b2,5d10,5x6,
5b2,6f9,6b2,VE=2.

8.66 Element 66 is Dyprosium or Dy

Dyprosium has 66 protons and 97 neutrons.

These are its isotopes which are stable or have a half-life over 1 year.

156Dy, 158Dy, 160Dy, 161Dy, 162Dy, 163Dy, 164Dy

154Dy = 3E6 y

This is its electron configuration using the new ring names.

66,Dy,1b2,2x6,2b2,3x6,3b2,4d10,4x6,4b2,5d10,5x6, 5b2,6f10,6b2,VE=2.

8.67 Element 67 is Holmium or Ho

Holmium has 67 protons and 98 neutrons.

These are its isotopes which are stable or have a half-life over 1 year.

165Ho

163Ho = 4570 y

This is its electron configuration using the new ring names.

67,Ho,1b2,2x6,2b2,3x6,3b2,4d10,4x6,4b2,5d10,5x6, 5b2,6f11,6b2,VE=2.

8.68 Element 68 is Erbium or Er

Erbium has 68 protons and 99 neutrons.

These are its isotopes which are stable or have a half-life over 1 year.

162Er, 164Er, 166Er, 167Er, 168Er, Er170

This is its electron configuration using the new ring names.

68,Er,1b2,2x6,2b2,3x6,3b2,4d10,4x6,4b2,5d10,5x6, 5b2,6f12,6b2,VE=2.

8.69 Element 69 is Thulmium or Tm

Thulmium has 69 protons and 100 neutrons.
These are its isotopes which are stable or have a half-life
over 1 year.

169Tm

171Tm = 1.92 y

This is its electron configuration using the new ring names.

69,Tm,1b2,2x6,2b2,3x6,3b2,4d10,4x6,4b2,5d10,
5x6,5b2,6f13,6b2,VE=2.

8.70 Element 70 is Ytterbium or Yb

Ytterbium has 70 protons and 103 neutrons.

These are its isotopes which are stable or have a half-life over 1 year.

168Yb, 170Yb, 171Yb, 172Yb, 173Yb,174Yb, 176Yb

This is its electron configuration using the new ring names.

70,Yb,1b2,2x6,2b2,3x6,3b2,4d10,4x6,4b2,5d10,5x6, 5b2,6f14,6b2,VE=2.

8.71 Element 71 is Lutetium or Lu

Lutetium has 71 protons and 104 neutrons.

These are its isotopes which are stable or have a half-life over 1 year.

175Lu

173Lu = 1.4 y
174Lu = 3.3 y
176Lu = 39E9 y

This is its electron configuration using the new ring names.

71,Lu,1b2,2x6,2b2,3x6,3b2,4d10,4x6,4b2,5d10,5x6,
5b2,6f14,6d1,6b2,VE=3.

8.72 Element 72 is Hafnium or Hf

Hafnium has 72 protons and 106 neutrons.

These are its isotopes which are stable or have a half-life over 1 year.

176Hf, 177Hf, 178Hf, 179Hf, 180Hf

182Hf = 39E9 y

This is its electron configuration using the new ring names.

72,Hf,1b2,2x6,2b2,3x6,3b2,4d10,4x6,4b2,5d10,5x6,5b2,6f14,6d2,6b2,VE=4.

8.73 Element 73 is Tantalum or Ta

Tantalum has 73 protons and 108 neutrons.

These are its isotopes which are stable or have a half-life over 1 year.

181Ta

This is its electron configuration using the new ring names.

73,Ta,1b2,2x6,2b2,3x6,3b2,4d10,4x6,4b2,5d10,5x6, 5b2,6f14,6d3,6b2,VE=5.

8.74 Element 74 is Tungsten or W

Tungsten has 74 protons and 110 neutrons.

These are its isotopes which are stable or have a half-life over 1 year.

182W, 183W, 184W, 186W

180W = 1.8E18 y

This is its electron configuration using the new ring names.

74,W,1b2,2x6,2b2,3x6,3b2,4d10,4x6,4b2,5d10,5x6,
5b2,6f14,6d4,6b2,VE=6.

8.75 Element 75 is Rhenium or Re

Rhenium has 75 protons and 111 neutrons.

These are its isotopes which are stable or have a half-life over 1 year.

184Re

186Re = 2E15 y

This is its electron configuration using the new ring names.

75,Re,1b2,2x6,2b2,3x6,3b2,4d10,4x6,4b2,5d10,5x6,
5b2,6f14,6d5,6b2,VE=7.

8.76 Element 76 is Osmium or Os

Osmium has 76 protons and 114 neutrons.

These are its isotopes which are stable or have a half-life over 1 year.

184Os, 187Os, 188Os, 189Os, 190Os, 192Os

186Os = 2E15 y
194Os = 6 y

This is its electron configuration using the new ring names.

76,Os,1b2,2x6,2b2,3x6,3b2,4d10,4x6,4b2,5d10,5x6, 5b2,6f14,6d6,6b2,VE=8.

8.77 Element 77 is Iridium or Ir

Iridium has 77 protons and 115 neutrons.

These are its isotopes which are stable or have a half-life over 1 year.

191Ir, 193Ir

This is its electron configuration using the new ring names.

77,Ir,1b2,2x6,2b2,3x6,3b2,4d10,4x6,4b2,5d10,5x6, 5b2,6f14,6d7,6b2,VE=9.

8.78 Element 78 is Platinum or Pt

Platinum has 78 protons and 117 neutrons.

These are its isotopes which are stable or have a half-life over 1 year.

192Pt, 194Pt, 195Pt, 196Pt, 198Pt

190Pt = 6.5E11 y
193Pt = 50 y

This is its electron configuration using the new ring names.

78,Pt,1b2,2x6,2b2,3x6,3b2,4d10,4x6,4b2,5d10,5x6, 5b2,6f14,6d9,6b1,VE=10.

This configuration has an anomaly from expected, with an electron moved between adjacent rings.

The change from expected: 6d8, 6b2 is 6d9,6b1.

8.79 Element 79 is Gold or Au

Gold has 79 protons and 118 neutrons.

These are its isotopes which are stable or have a half-life over 1 year.

197Au

This is its electron configuration using the new ring names.

79,Au,1b2,2x6,2b2,3x6,3b2,4d10,4x6,4b2,5d10,5x6, 5b2,6f14,6d10,6b1,VE=1

This configuration has an anomaly from expected, with an electron moved between adjacent rings.

The change from expected: 6d9, 6b2 is 6d10,6b1.

8.80 Element 80 is Mercury or Hg

Mercury has 80 protons and 121 neutrons.

These are its isotopes which are stable or have a half-life over 1 year.

196Hg, 198Hg, 198Hg, 199Hg, 200Hg, 201Hg, 202Hg, 204Hg

This is its electron configuration using the new ring names.

80,Hg,1b2,2x6,2b2,3x6,3b2,4d10,4x6,4b2,5d10,5x6, 5b2,6f14,6d10,6b2,VE=2.

8.81 Element 81 is Thallium or Tl

Thallium has 81 protons and 123 neutrons.

These are its isotopes which are stable or have a half-life over 1 year.

203Tl, 205Tl

204Tl = 3.8 y

This is its electron configuration using the new ring names.

81,Tl,1b2,2x6,2b2,3x6,3b2,4d10,4x6,4b2,5d10,5x6,
5b2,6f14,6d10,6x1,6b2,VE=3.

8.82 Element 82 is Lead or Pb

Lead has 82 protons and 125 neutrons.

These are its isotopes which are stable or have a half-life over 1 year.

204Pb, 206Pb, 207Pb, 208Pb
210Pb = 22.3 y

This is its electron configuration using the new ring names.

82,Pb,1b2,2x6,2b2,3x6,3b2,4d10,4x6,4b2,5d10,5x6,
5b2,6f14,6d10,6x2,6b2,VE=4.

8.83 Element 83 is Bismuth or Bi

Bismuth has 83 protons and 126 neutrons.

These are its isotopes which are stable or have a half-life over 1 year.

20Bi = 32.9 y
209Bi = 2E19 y (longest life)

This is its electron configuration using the new ring names.

83,Bi,1b2,2x6,2b2,3x6,3b2,4d10,4x6,4b2,5d10,5x6,
5b2,6f14,6d10,6x3,6b2,VE=5.

8.84 Element 84 is Polonium or Po

Polonium has 84 protons and 125 neutrons.

These are its isotopes which are stable or have a half-life over 1 year.

208Po = 2.9 y
209Po = 25.2 y (longest life)

This is its electron configuration using the new ring names.

84,Po,1b2,2x6,2b2,3x6,3b2,4d10,4x6,4b2,5d10,5x6,
5b2,6f14,6d10,6x4,6b2,VE=6.

8.85 Element 85 is Astatine or At

Astatine has 85 protons and 125 neutrons.

These are its isotopes which are stable or have a half-life over 1 year.

210At = 8.1 h (longest life)

This is its electron configuration using the new ring names.

85,At,1b2,2x6,2b2,3x6,3b2,4d10,4x6,4b2,5d10,5x6, 5b2,6f14,6d10,6x5,6b2,VE=7.

8.86 Element 86 is Radon or Rn

Radon has 86 protons and 136 neutrons.

These are its isotopes which are stable or have a half-life over 1 year.

222Rn = 3.8 d (longest life)

This is its electron configuration using the new ring names.

86,Rn,1b2,2x6,2b2,3x6,3b2,4d10,4x6,4b2,5d10,5x6, 5b2,6f14,6d10,6x6,6b2,VE=8.

8.87 Element 87 is Francium or Fr

Francium has 87 protons and 136 neutrons.

These are its isotopes which are stable or have a half-life over 1 year.

223Fr = 22 m (longest life)

This is its electron configuration using the new ring names.

87,Fr,1b2,2x6,2b2,3x6,3b2,4d10,4x6,4b2,5d10,5x6,
5b2,6f14,6d10,6x6,6b2,7b1,VE=1.

8.88 Element 88 is Radium or Ra

Radium has 88 protons and 138 neutrons.

These are its isotopes which are stable or have a half-life over 1 year.

226Ra = 1600 y (longest life)

This is its electron configuration using the new ring names.

88,Ra,1b2,2x6,2b2,3x6,3b2,4d10,4x6,4b2,5d10,5x6, 5b2,6f14,6d10,6x6,6b2,7b2,VE=2.

8.89 Element 89 is Actinium or Ac

Actinium has 89 protons and 138 neutrons.

These are its isotopes which are stable or have a half-life over 1 year.

227Ac = 21.8 y (longest life)

This is its electron configuration using the new ring names.

89,Ac,1b2,2x6,2b2,3x6,3b2,4d10,4x6,4b2,5d10,5x6,
5b2,6f14,6d10,6x6,6b2,7d1,7b2,VE=3.

This configuration has an anomaly from expected, with an electron moved between adjacent rings.

The change from expected: 7f1, 7d0 is 7f0,7d1.

8.90 Element 90 is Thorium or Th

Thorium has 90 protons and 142 neutrons.

These are its isotopes which are stable or have a half-life over 1 year.

232Th = 1.4E10 y (longest life)

This is its electron configuration using the new ring names.

90,Th,1b2,2x6,2b2,3x6,3b2,4d10,4x6,4b2,5d10,5x6, 5b2,6f14,6d10,6x6,6b2,7d2,7b2,VE=4.

8.91 Element 91 is Protactinium or Pa

Protactinium has 91 protons and 140 neutrons.

These are its isotopes which are stable or have a half-life over 1 year.

231Pa = 3.3E4 y (longest life)

This is its electron configuration using the new ring names.

91,Pa,1b2,2x6,2b2,3x6,3b2,4d10,4x6,4b2,5d10,5x6,
5b2,6f14,6d10,6x6,6b2,7f2,7d1,7b2,VE=3.

8.92 Element 92 is Uranium or U

Uranium has 92 protons and 146 neutrons.

These are its isotopes which are stable or have a half-life over 1 year.

238U = 4.5E9 y (longest life)

This is its electron configuration using the new ring names.

92,U,1b2,2x6,2b2,3x6,3b2,4d10,4x6,4b2,5d10,5x6,
5b2,6f14,6d10,6x6,6b2,7f3,7d1,7b2,VE=3.

8.93 Element 93 is Neptunium or Np

Neptunium has 93 protons and 144 neutrons.

These are its isotopes which are stable or have a half-life over 1 year.

237Np = 2.21E6 y (longest life)

This is its electron configuration using the new ring names.

93,Np,1b2,2x6,2b2,3x6,3b2,4d10,4x6,4b2,5d10,5x6,
5b2,6f14,6d10,6x6,6b2,7f4,7d1,7b2,VE=3.

8.94 Element 94 is Plutonium or Pu

Plutonium has 94 protons and 150 neutrons.

These are its isotopes which are stable or have a half-life over 1 year.

244Pu = 8E7 y (longest life)

This is its electron configuration using the new ring names.

94,Pu,1b2,2x6,2b2,3x6,3b2,4d10,4x6,4b2,5d10,5x6,
5b2,6f14,6d10,6x6,6b2,7f6,7b2,VE=2.

8.95 Element 95 is Americium or Am

Americium has 95 protons and 148 neutrons.

These are its isotopes which are stable or have a half-life over 1 year.

243Am = 7370 y (longest life)

This is its electron configuration using the new ring names.

95,Am,1b2,2x6,2b2,3x6,3b2,4d10,4x6,4b2,5d10,5x6, 5b2,6f14,6d10,6x6,6b2,7f7,7b2,VE=2.

8.96 Element 96 is Curium or Cm

Curium has 96 protons and 151 neutrons.

These are its isotopes which are stable or have a half-life over 1 year.

247Cm = 1.6E7 y (longest life)

This is its electron configuration using the new ring names.

96,Cm,1b2,2x6,2b2,3x6,3b2,4d10,4x6,4b2,5d10,
5x6,5b2,6f14,6d10,6x6,6b2,7f7,7d1,7b2,VE=3.

8.97 Element 97 is Berkelium or Bk

Berkelium has 97 protons and 150 neutrons.

These are its isotopes which are stable or have a half-life over 1 year.

247Bk = 1.4E3 y (longest life)

This is its electron configuration using the new ring names.

997,Bk,1b2,2x6,2b2,3x6,3b2,4d10,4x6,4b2,5d10,
5x6,5b2,6f14,6d10,6x6,6b2,7f8,7d1,7b2,VE=3.

This configuration has an anomaly from expected, with an electron moved between adjacent rings.

The change from expected: 7f1, 7d0 is 7f0,7d1.

8.98 Element 98 is Californium or Cf

Californium has 98 protons and 153 neutrons.
These are its isotopes which are stable or have a half-life
over 1 year.

251Cf = 900 y (longest life)

This is its electron configuration using the new ring names.

98,Cf,1b2,2x6,2b2,3x6,3b2,4d10,4x6,4b2,5d10,
5x6,5b2,6f14,6d10,6x6,6b2,7f10,7b2,VE=2.

8.99 Element 99 is Einsteinium or Es

Einsteinium has 99 protons and 153 neutrons.

These are its isotopes which are stable or have a half-life over 1 year.

252Es = 471.7 y (longest life)

This is its electron configuration using the new ring names.

99,Es,1b2,2x6,2b2,3x6,3b2,4d10,4x6,4b2,5d10,
5x6,5b2,6f14,6d10,6x6,6b2,7f11,7b2,VE=2.

8.100 Element 100 is Fermium or Fm

Fermium has 100 protons and 157 neutrons.

These are its isotopes which are stable or have a half-life over 1 year.

257Fm = 100.5 d (longest life)

This is its electron configuration using the new ring names.

100,Fm,1b2,2x6,2b2,3x6,3b2,4d10,4x6,4b2,
5d10,5x6,5b2,6f14,6d10,6x6,6b2,7f12,7b2,VE=2.

8.101 Element 101 is Mendelevium or Md

Mendelevium has 101 protons and 157 neutrons.

These are its isotopes which are stable or have a half-life over 1 year.

258Md = 51.5 d (longest life)

This is its electron configuration using the new ring names.

101,Md,1b2,2x6,2b2,3x6,3b2,4d10,4x6,
4b2,5d10,5x6,5b2,6f14,6d10,6x6,6b2,7f13,7b2,VE=2.

8.102 Element 102 is Nobelium or No

Nobelium has 102 protons and 157 neutrons.

These are its isotopes which are stable or have a half-life over 1 year.

259No = 58 m (longest life)

This is its electron configuration using the new ring names.

102,No,1b2,2x6,2b2,3x6,3b2,4d10,4x6,
4b2,5d10,5x6,5b2,6f14,6d10,6x6,6b2,7f14,7b2,VE=2.

8.103 Element 103 is Lawrencium or Lr

Lawrencium has 103 protons and 159 neutrons.

These are its isotopes which are stable or have a half-life over 1 year.

262Lr = 216 m (longest life)

This is its electron configuration using the new ring names.

103,Lr,1b2,2x6,2b2,3x6,3b2,4d10,4x6,
4b2,5d10,5x6,5b2,6f14,6d10,6x6,6b2,7f14,7x1,7b2,VE=3.

This configuration has an anomaly from expected, with an electron moved between adjacent rings.

The change from expected: 7d1, 7x0 is 7d0,7x1.

8.104 Element 104 is Rutherfordium or Rf

Rutherfordium has 104 protons and 157 neutrons.
These are its isotopes which are stable or have a half-life over 1 year.

261Rf = 68 s (too short)

This is its electron configuration using the new ring names.

104,Rf,1b2,2x6,2b2,3x6,3b2,4d10,4x6,
4b2,5d10,5x6,5b2,6f14,6d10,6x6,6b2,7f14,7d2,7b2,VE=2.

8.105 Element 105 is Dubnium or Db

Dubnium has 105 protons and 157 neutrons.

These are its isotopes which are stable or have a half-life over 1 year.

262Db = 35 s (too short)

This is its electron configuration using the new ring names.

105,Db,1b2,2x6,2b2,3x6,3b2,4d10,4x6,
4b2,5d10,5x6,5b2,6f14,6d10,6x6,6b2,7f14,7d3,7b2,VE=2.

8.106 Element 106 is Seaborgium or Sg

Seaborgium has 106 protons and 157 neutrons.

These are its isotopes which are stable or have a half-life over 1 year.

266Sg = 0.76 s (too short)

This is its electron configuration using the new ring names.

106,Sg,1b2,2x6,2b2,3x6,3b2,4d10,4x6,
4b2,5d10,5x6,5b2,6f14,6d10,6x6,6b2,7f14,7d4,7b2,VE=2.

8.107 Element 107 is Borhrium or Bh

Borhrium has 107 protons and 157 neutrons.

These are its isotopes which are stable or have a half-life over 1 year.

2614Bh = 1.1 s (too short)

This is its electron configuration using the new ring names.

107,Bh,1b2,2x6,2b2,3x6,3b2,4d10,4x6,
4b2,5d10,5x6,5b2,6f14,6d10,6x6,6b2,7f14,7d5, 7b2,VE=2.

8.108 Element 108 is Hassium or Hs

Hassium has 108 protons and 159 neutrons.

Hassium has no isotopes which are stable or have a half-life over 1 year.

This is its electron configuration using the new ring names.

108,Hs,1b2,2x6,2b2,3x6,3b2,4d10,
4x6,4b2,5d10,5x6,5b2,6f14,6d10,6x6,6b2,7f14,
7d6,7b2,VE=2.

8.109 Element 109 is Meiterium or Mt

Meiterium has 109 protons and 159 neutrons.

Meiterium has no isotopes which are stable or have a half-life over 1 year.

This is its electron configuration using the new ring names.

109,Mt,1b2,2x6,2b2,3x6,3b2,4d10,
4x6,4b2,5d10,5x6,5b2,6f14,6d10,6x6,6b2,
7f14,7d7,7b2,VE=2.

8.110 Element 110 is Damstadtium or Ds

Damstadtium has 110 protons and 152 neutrons.

Damstadtium has no isotopes which are stable or have a half-life over 1 year.

This is its electron configuration using the new ring names.

110,Ds,1b2,2x6,2b2,3x6,3b2,4d10,
4x6,4b2,5d10,5x6,5b2,6f14,6d10,
6x6,6b2,7f14,7d8,7b2,VE=2.

8.111 Element 111 is Roehtgenium Rg

Roehtgenium has 111 protons and 161 neutrons.

Roehtgenium has no isotopes which are stable or have a half-life over 1 year.

This is its electron configuration using the new ring names.

111,Rg,1b2,2x6,2b2,3x6,3b2,4d10,
4x6,4b2,5d10,5x6,5b2,6f14,6d10,6x6,6b2,
7f14,7d10,7b1,VE=1.

This configuration has an anomaly from expected, with an electron moved between adjacent rings.

The change from expected: 7d9, 7b2 is 7d10,7b2.

8.112 Element 112 is Copernicium or Cn

Copernicium has 112 protons and 173 neutrons.

Copernicium has no isotopes which are stable or have a half-life over 1 year.

This is its electron configuration using the new ring names.

1112,Cn,1b2,2x6,2b2,3x6,3b2,4d10,
4x6,4b2,5d10,5x6,5b2,6f14,6d10,6x6,6b2,
7f14,7d10,7b2,VE=2.

8.113 Element 113 is Nihonium or Nh

Nihonium has 113 protons and 173 neutrons.

Nihonium has no isotopes which are stable or have a half-life over 1 year.

This is its electron configuration using the new ring names.

113,Nh,1b2,2x6,2b2,3x6,3b2,4d10,
4x6,4b2,5d10,5x6,5b2,6f14,6d10,6x6,
6b2,7f14,7d10,7x1,7b2,VE=3.

8.114 Element 114 is Flerovium or Fl

Flerovium has 114 protons and 175 neutrons.

Flerovium has no isotopes which are stable or have a half-life over 1 year.

This is its electron configuration using the new ring names.

114,Fl,1b2,2x6,2b2,3x6,3b2,4d10,
4x6,4b2,5d10,5x6,5b2,6f14,6d10,
6x6,6b2,7f14,7d10,7x2,7b2,VE=4.

8.115 Element 115 is Muscovium or Mc

Muscovium has 115 protons and 173 neutrons.

Muscovium has no isotopes which are stable or have a half-life over 1 year.

This is its electron configuration using the new ring names.

115,Mc,1b2,2x6,2b2,3x6,3b2,4d10,
4x6,4b2,5d10,5x6,5b2,6f14,6d10,
6x6,6b2,7f14,7d10,7x3,7b2,VE=5.

8.116 Element 116 is Livermorium or Lv

Livermorium has 116 protons and 177 neutrons.

Livermorium has no isotopes which are stable or have a half-life over 1 year.

This is its electron configuration using the new ring names.

116,Lv,1b2,2x6,2b2,3x6,3b2,4d10,
4x6,4b2,5d10,5x6,5b2,6f14,6d10,
6x6,6b2,7f14,7d10,7x4,7b2,VE=6.

8.117 Element 117 is Tennessine or Ts

Tennessine has 117 protons and 177 neutrons.

Tennessine has no isotopes which are stable or have a half-life over 1 year.

This is its electron configuration using the new ring names.

117,Ts,1b2,2x6,2b2,3x6,3b2,4d10,
4x6,4b2,5d10,5x6,5b2,6f14,6d10,
6x6,6b2,7f14,7d10,7x5,7b2,VE=7.

8.118 Element 118 is Oganesson or Og

Oganesson has 118 protons and 176 neutrons.

Oganesson has no isotopes which are stable or have a half-life over 1 year.

This is its electron configuration using the new ring names.

118,Og,1b2,2x6,2b2,3x6,3b2,4d10,
4x6,4b2,5d10,5x6,5b2,6f14,6d10,
6x6,6b2,7f14,7d10,7x6,7b2,VE=8.

9 Atomic Model

The Standard Model is quite flawed when treating electrons as waves. They are real particles, having a negative charge. They are subject to Coulomb's force between charged particles. In an atom, there are electrons in the nucleus, helping maintain stability with protons in contact or proximity.

Electrons are in something of a cloud around the nucleus, affected by both the positively charged nucleus and the other negatively charged electrons. There is a balance among these individual forces within an atom.

This author is proposing circular orbits for the electrons because that explains the atomic radius measurements and their predictability suits the use of chemistry and its use of the valence electrons in the outer ring.

All the evidence points to circular rings of electrons around the nucleus. Each element has a defined electron configuration which chemists have used since about 1949.

The irony is the Bohr model was nearly accepted though having circular obits.

If there is a requirement the atomic model must be compatible with relativity, this author will take exception to such a dubious requirement. The author's first 5 books had various justifications for removing relativity from physics. There is no need for integrating the context of a special observer, subject only to gravitational fields, in an atomic model.

9.1 Bohr atomic model

Excerpt from Wikipedia:

In 1913, Niels Bohr identified the correspondence principle and used it to formulate a model of the hydrogen atom which explained the line spectrum. In the next few years Arnold Sommerfeld extended the quantum rule to arbitrary integrable systems making use of the principle of adiabatic invariance of the quantum numbers introduced by Lorentz and Einstein. Sommerfeld made a crucial contribution by quantizing the z-component of the angular momentum, which in the old quantum era was called space quantization (Richtungsquantelung). This allowed the orbits of the electron to be ellipses instead of circles, and introduced the concept of quantum degeneracy. The theory would have correctly explained the Zeeman effect, except for the issue of electron spin. Sommerfeld's model was much closer to the modern quantum mechanical picture than Bohr's.

(Excerpt end)

Observation:

Sommerfield "extended the quantum rule making use of quantum numbers. This allowed the orbits of the electron to be ellipses instead of circles."

This used "allowed" but not "restricted"

9.1 Lewis Dot structure

In 1916, Gilbert Lewis proposed a dot structure to represent the number of electrons in the circular valence ring of each element.

Excerpt from Wikipedia:

Lewis structures, also known as Lewis dot formulas, Lewis dot structures, electron dot structures, or Lewis electron dot structures (LEDS), are diagrams that show the bonding between atoms of a molecule and the lone pairs of electrons that may exist in the molecule. A Lewis structure can be drawn for any covalently bonded molecule, as well as coordination compounds. The Lewis structure was named after Gilbert N. Lewis, who introduced it in his 1916 article The Atom and the Molecule. Lewis structures extend the concept of the electron dot diagram by adding lines between atoms to represent shared pairs in a chemical bond.

(Excerpt end)

Observation:

For over 100 years, chemistry has relied on the consistent behavior in an atom's electron configuration.

There is never a margin of error to accommodate the uncertainty of electrons being in their predicted rings at the moment when required.

Lewis structures are a simple representation of the valence ring, but it is so useful because the electrons are always where expected.

The supposed uncertainty with electrons which arose with quantum mechanics is only theoretical. That uncertainty does not exist in real atoms.

Simple tools like Lewis dot structures remain valid in a practical atomic model because an electron configuration is still predictable, like in 1916.

9.2 Valence bond theory

The Bohr model needed improvements, which came with valence bond theory.

Excerpt from Wikipedia:

According to this theory a covalent bond is formed between two atoms by the overlap of half filled valence atomic orbitals of each atom containing one unpaired electron. A valence bond structure is similar to a Lewis structure, but where a single Lewis structure cannot be written, several valence bond structures are used. Each of these VB structures represents a specific Lewis structure. This combination of valence bond structures is the main point of resonance theory. Valence bond theory considers that the overlapping atomic orbitals of the participating atoms form a chemical bond. Because of the overlapping, it is most probable that electrons should be in the bond region. Valence bond theory views bonds as weakly coupled orbitals (small overlap). Valence bond theory is typically easier to employ in ground state molecules. The core orbitals and electrons remain essentially unchanged during the formation of bonds.

(Excerpt end)

Observation:

The phrase "it is most probable" is apparently needed to accommodate atoms which are not in ground state,

9.3 Future elements

Currently, mankind has identified 118 distinct elements.

It is possible heavier elements can be created. New elements are being created in laboratories using particle accelerators rather than a process of fusion with pressure for compressing particles into an existing nucleus.

Uncontrolled high velocity particle collisions are unable to create stable isotopes, by observation. Most of the heavier isotopes quickly decay.

Supposedly, element 118 was discovered in debris, in Roswell, New Mexico, in 1947. This detection of a stable 118 which is not stable here in laboratories implies a process of transmutation (described in section 4), not collisions, could create heavier isotopes having a longer life before decay.

We have never used fusion for compression, using extremely high pressure with electromagnetic forces, when attempting an element's creation. A collision is not compression. As noted earlier, a plasma pinch has the potential for sufficient force of compression for fusion. As a reference, initial controlled fusion reactors used a pinch mechanism, not a nuclei collider.

If elements after 118 are eventually created, they will require names.

This author has a suggestion using hexadecimal digits rather than being limited to the letters of the English alphabet.

The simple naming convention of future elements is the capital letter X followed by 2 hex digits, where X represents 100 and the decimal value of the 2 hex digits is added to 100. This scheme covers to 355 for 1FF.

X00 to X12 are already named, from elements 100 to 118. Therefore, the possible names XA through XF are taken. A naming conflict with Xe for Xenon is averted.

X13 is a future atom 119.

X32 is future atom 150 completes the next cycle of bfdx electron rings. They fill at 150, from 118+32.

The current element naming convention is -ium for groups 1-6, -ine for group 7 and something like -enon for group 8 (with the inert gas elements. Each cycle spans rings of bfdx, with 2 and 6 at the ends. A new cycle begins with 1.

X11 was named Tennessine, with –ine just like fluorine and iodine.

With this naming convention, X13 through X30 are named Xhhium, with hh for 2 hex digits. Therefore, X13 is named X13ium. X14 is named X14ium. X31 is named X31ine as another group 17 element. X32 is named something like X32enon, as another group 18 element.

9.4 Ionization energy

This atomic behavior is related to valence and has notable exceptions.

Excerpt from Wikipedia:

In physics and chemistry, ionization energy is the minimum amount of energy required to remove the most loosely bound electron of an isolated neutral gaseous atom or molecule. It is quantitatively expressed as
$$X(g) + energy \rightarrow X+(g) + e-$$
where X is any atom or molecule, X+ is the ion with one electron removed, and e- is the removed electron. This is generally an endothermic process. As a rule, the closer the outermost electrons to the nucleus of the atom, the higher the atom's ionization energy.

The sciences of physics and chemistry use different units for ionization energy. In physics, the unit is the amount of energy required to remove a single electron from a single atom or molecule, expressed as electronvolts. In chemistry, the unit is the amount of energy required for all of the atoms in a mole of substance to lose one electron each: molar ionization energy or approximately enthalpy, expressed as kilojoules per mole (kJ/mol) or kilocalories per mole (kcal/mol).

Comparison of ionization energies of atoms in the periodic table reveals two periodic trends which follow the rules of Coulombic attraction:

Ionization energy generally increases from left to right within a given period (that is, row).

Ionization energy generally decreases from top to bottom in a given group (that is, column).

The latter trend results from the outer electron shell being progressively farther from the nucleus, with the addition of one inner shell per row as one moves down the column.

Ionization energy of atoms, denoted Ei, is measured by finding the minimal energy of light quanta (photons) or electrons accelerated to a known energy that will kick out the least bound atomic electrons. The measurement is performed in the gas phase on single atoms. While only noble gases occur as monoatomic gases, other gases can be split into single atoms. Also, many solid elements can be heated and vaporized into single atoms. Monoatomic vapor is contained in a previously evacuated tube that has two parallel electrodes connected to a voltage source. The ionizing excitation is introduced through the walls of the tube or produced within.

(Excerpt end)

Observation: The topic has extensive descriptions of notable exceptions to the general rules.

The measurements confirm the electrons must be in rings and not behaving as waves based on probabilities.

The section 3 Data Sets identifies the worksheet having the ionization energies. These data are presented in the following charts.

9.4.1 First ionization energy

This is the energy, measured in electronvolts (as described in the earlier excerpt, required for one electron to be ejected by the atom.

The atom becomes an ion because now there is 1 more net positive charge in the nucleus than the total negative charges in the remaining electrons.

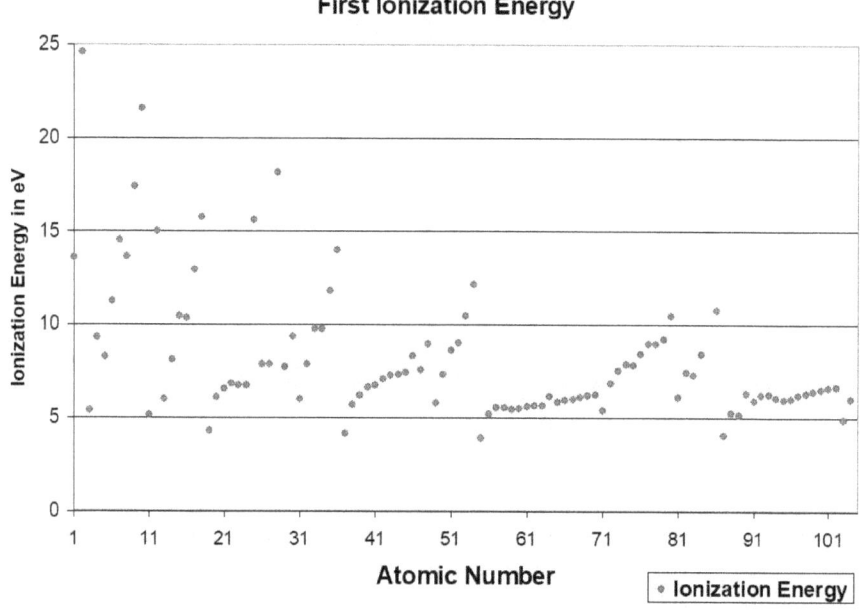

First Ionization Energy

9.4.2 Second ionization energy

This is the energy, measured in electronvolts required for the second electron to be ejected by the atom.

The atom is an ion because now there are 2 more net positive charges in the nucleus than the total negative charges in the remaining electrons.

Second Ionization Energy

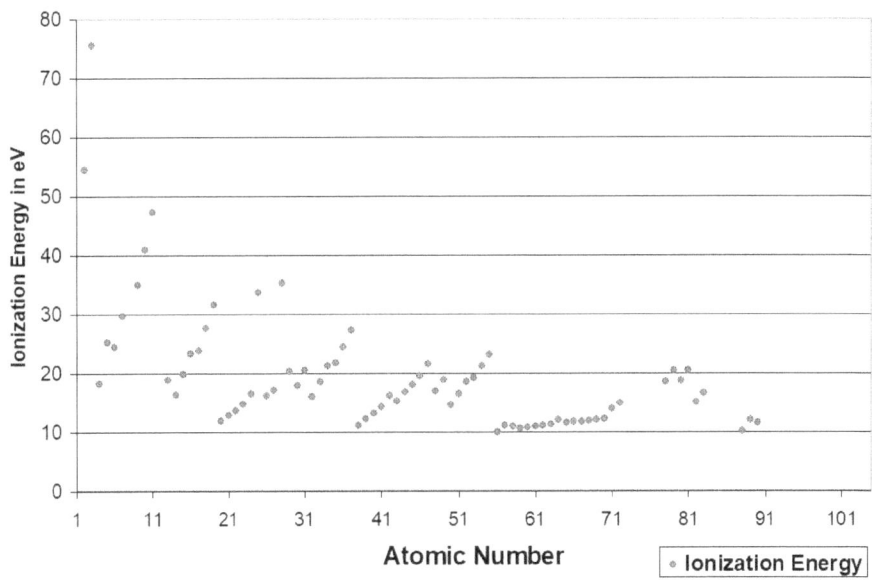

9.4.3 Third ionization energy

This is the energy, measured in electronvolts required for the third electron to be ejected by the atom.

The atom is an ion because now there are 3 more net positive charges in the nucleus than the total negative charges in the remaining electrons.

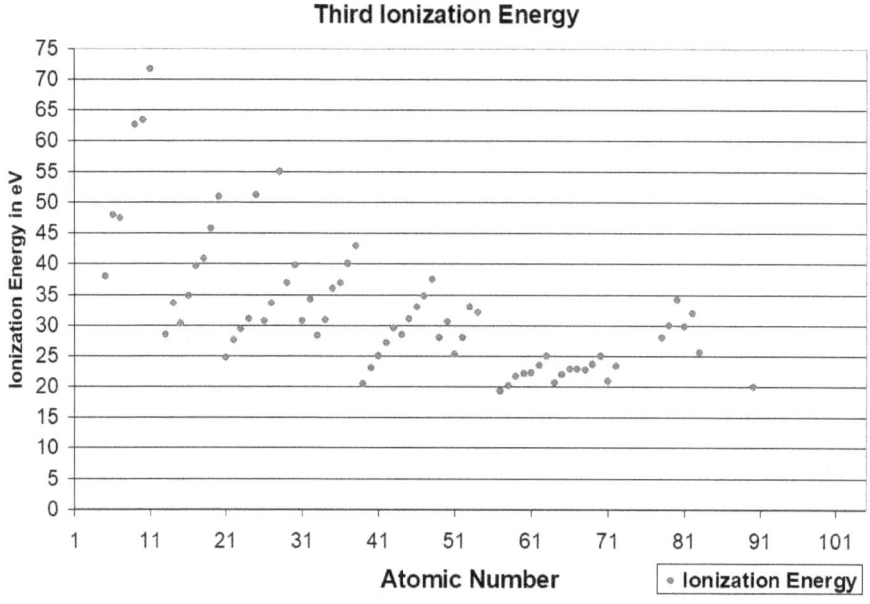

Third Ionization Energy

9.4.3 Conclusion from the published ionization energy values.

When quantum mechanics disturbed the Bohr model having circular orbits, the result was the unverified theory of an electron having an uncertain position. This uncertainty was described as a wave function which was claimed to define the probability of the electron being at a particular location in its noncircular path.

When there is only a probability of an electron being in the expected location, then its opposite must exist. There must be the corresponding probability the electron will not be in the required location at the moment of the measurement. That unexpected misposition must result in a different measurement.

With the proposed wave function behavior, then every measurement must have a stated margin of error to accommodate the stated probability for when an electron is not in its required position within its expected circular ring.

The fact none of these stated energy values has the required margin of error is confirmation electrons are predictable and the proposed wave function for a probability does not exist in real atoms. The electron's wave function must be removed and cannot be part of a practical atomic model.

9.5 Electron as Wave

There is at least one experiment claiming an electron is a particle with a wave behavior. One of them is the electron diffraction experiment using a double-slit technique, like used with light. James Clerk Maxwell thoroughly explained light is a wave because light is the propagation of perpendicular, synchronized, electric and magnetic fields. The rate of oscillation of these fields can be measured as a frequency or wave length.

It is this author's position those double-slit experiments are never with electrons in an atom's predicted configuration so it is difficult to compare the 2 unrelated contexts and apply one observation in only one context to the other context,

An electron gun shooting a stream of electrons is technically generating a plasma filament. Charged particles are correctly called plasma. Plasma has unique behaviors, including self–organizing into filaments. It is possible the experiment failed to account for plasma behaviors.

Excerpt from Wikipedia:

An important version of this experiment involves single particles. Sending particles through a double-slit apparatus one at a time results in single particles appearing on the screen, as expected. Remarkably, however, an interference pattern emerges when these particles are allowed to build up one by one. This demonstrates the wave–particle.

An important version of this experiment involves single particles. Sending particles through a double-slit apparatus one at a time results in single particles appearing on the screen, as expected. Remarkably, however, an interference pattern emerges when these particles are allowed to build up one by one. This demonstrates the wave–particle duality, which states that all matter exhibits both wave and particle properties: the particle is measured as a single pulse at a single position, while the wave describes the probability of absorbing the particle at a specific place on the screen. This phenomenon has been shown to occur with photons, electrons, atoms and even some molecules, including buckyballs.

The probability of detection is the square of the amplitude of the wave and can be calculated with classical waves. Ever since the origination of quantum mechanics, some theorists have searched for ways to incorporate additional determinants or "hidden variables" that, were they to become known, would account for the location of each individual impact with the target.
This demonstrates the wave-particle duality, which states that all matter exhibits both wave and particle properties: the particle is measured as a single pulse at a single position, while the wave describes the probability of absorbing the particle at a specific place on the screen. This phenomenon has been shown to occur with photons, electrons, atoms and even some molecules, including buckyballs.
The probability of detection is the square of the amplitude of the wave and can be calculated with classical waves.

Ever since the origination of quantum mechanics, some theorists have searched for ways to incorporate additional determinants or "hidden variables" that, were they to become known, would account for the location of each individual impact with the target.

(Excerpt end)

Observation:

The analysis must be about only electrons because they have the claimed wave behavior.

This is poor description of an actual experiment. There are frequent diversions. Every reference to a thought experiment should be removed. They are irrelevant. Every reference to a photon should be removed. Light is a wave so it can cause the diffraction pattern with no question or further testing.
Calculations are mentioned but with no details, including their purpose, critical assumptions equations, factors, and results.

There is a search for "hidden variables" but most the critical variables in the experiment are not provided, so nearly all are hidden from analysis:
a) what range of slit width was used?
b) were the slit edges flat, round or chamfered?
c) what range of electron velocity was used?
d) what range of particle ejection frequency was used?
e) what range of particle quantity was used?
f) what range of alignment of gun to center of slit was used?
g) what range of lateral velocity of the gun was used?
h) which materials were used for the slits?

For an experiment creating the foundation of quantum mechanics, all the limits of every parameter should have been tested, especially many slit widths, as many as practical. There is no mention of testing such a critical variable of this experiment.

Having so little detail implies this is only a meaningless thought experiment for theoretical physicists who do not perform actual experiments. Those require a methodical approach enabling others to duplicate the test, to either confirm or deny any conclusions being drawn. Theories are nearly worthless until their predictions are tested.

9.6 Electron Particle Position Certainty

Section 5 Gather Electrons listed the radius of every electron for most of the 118 elements.
The electrons moving around the nucleus in a ring have adequate spacing. This is definitely not a linear beam of electrons and there should be no expectation for the ring to behave just like the particles are in a line.

When a new ring begins, there are these behaviors:

a) b ring: its radius is larger than the just completed ring.

The amount of the increase depends on the rings to be filled inside this b ring.

As the f, d, and x rings fill inside this outer b ring, the outer ring tends to decrease on size, probably because the filling of the f, d, and x rings is driven the increasing positive change in the nucleus from the increasing Z, the atomic number which identifies the element and its electron configuration.

9.6 Quantum Defect

When this book was conceived, quantum defect was a possible topic. After a little research, the supposed defect lacks enough detail to justify an effort with analysis.

Excerpt from Wikipedia:

The term quantum defect refers to two concepts: energy loss in lasers and energy levels in alkali elements. Both deal with quantum systems where matter interacts with light.
In laser science, the term "quantum defect" refers to the fact that the energy of a pump photon is generally higher than that of a signal photon (photon of the output radiation). The energy difference is lost to heat, which may carry away the excess entropy delivered by the multimode incoherent pump.

(Excerpt end)

Observation:

The term pump photon must refer to a light source.
Incoherent must refer to a pulsing light source.
Photon refers to a wave length of light. Light is electromagnetic energy emitted from a source having the necessary electric and magnetic fields to initiate propagation.
The description uses "generally higher" but with no data justifying this vague phrase. There is only a formula which lacks enough data to justify its usefulness. Apparently some energy is lost to heat in the object absorbing the light. This is a common event in thermodynamics.

No values are provided as an example, like detailed description of the original energy and the amount of heat or thermal energy measured.

Thermodynamics requires energy to be conserved. The description does not imply energy from the laser wave length is lost but rather some was transformed into thermal energy when absorbed. The word photon is used but light carries energy in a wave length of electromagnetic radiation. There is no photon. It is just the name of a quasi-particle used as an abstraction of one wavelength of light. If there is some variable amount of energy being transformed, then this cannot be a quantized and invoking a photon is not appropriate when the energy in the process is across a continuum, not in precise, predictable increments, or a quantum,

Particle physics has an atomic behavior called atomic mass defect where the measured mass of an atom does not match the sum of its particles. The author's previous book, Practical Particle Physics, thoroughly explained this supposed defect is caused by a proton losing some of its measured mass when the particle's volume is reduced by the compression during fusion into the nucleus. There is no conversion of mass to binding energy as proposed for a possible, but wrong, explanation of the difference in measured mass.

The description of quantum defect lacks the necessary details which enabled mass defect getting a practical explanation. Quantum defect is not worthy of further mention in this book.

10 Light

There are several interactions between light or electromagnetic radiation and atoms or matter.

This section is adapted from this topic in the author's book, Observing Our Universe.

That book offers much more detail than necessary for this book about the atomic model. Only the basics are here.

10.1 Light and wavelengths

A spectrum is the entire range of wavelengths in electromagnetic radiation where light is the visible range. The ultraviolet and infrared ranges are not visible to the human eye but they are in the Sun's radiation. Because this radiation can come from sources spanning beyond the visible range and for simplicity, the word light is often used for the entire spectrum, including those frequency ranges not visible.

Electromagnetic radiation is the propagation of synchronized, perpendicular electric and magnetic fields. The propagation has a defined rate of oscillation measured as either a frequency or a wavelength. The wavelength is usually measured in either nanometers (10^{-9} m) or Angstroms (10^{-10} m or 0.1 nm). The velocity of this propagation has been measured in a vacuum using our standard definition for time and this measured value is called the constant c. This measurement also defined the standard unit of 1 meter. The velocity of propagation is reduced in a medium, defined by the medium's diffraction index.

Light transmits energy proportional to its frequency so the constant c appears in some physics equations involving energy.

Quantum physics defined a theoretical particle called a photon to refer to a single wavelength.
In this section, wavelength is used because a spectrum analysis uses specific numerical values. Using the word photon instead of wavelength only introduces possible confusion when the radiation is a continuum of energy having no discrete values.

10.2 Fraunhofer Lines

This description provides background for many terms and their use in a spectrum analysis.

Excerpt from Wikipedia:

In 1814, Fraunhofer independently rediscovered the [dark] lines and began to systematically study and measure the wavelengths where these features are observed. He mapped over 570 lines.

About 45 years later Kirchhoff and Bunsen noticed that several Fraunhofer lines coincide with characteristic emission lines identified in the spectra of heated elements. It was correctly deduced that dark lines in the solar spectrum are caused by absorption by chemical elements in the solar atmosphere. Some of the observed features were identified as telluric lines originating from absorption by oxygen molecules in the Earth's atmosphere.

Because of their well–defined wavelengths, Fraunhofer lines are often used to characterize the refractive index and dispersion properties of optical materials.

(Excerpt end)

10.3 Atom's characteristic wavelengths

Calcium and hydrogen are the most frequently observed atoms in the spectrum of a distant galaxy or quasar.

The calcium atom is important because a galaxy can have its ion's pair of calcium absorption lines at 3934 and 3969 Angstroms in its spectrum when a calcium ion is in the line of sight to the galaxy. A red or blue shift of this pair of lines indicates the relative velocity of the ion. The neutral calcium atom has a different pair of wavelengths. Nearly all matter in the universe is plasma, or it has an electrical charge. That includes electrons (-), protons(+), and ions (+) which are atoms having lost one or more electrons.

Hydrogen is the most common element in the universe; it is also the simplest having only one proton and one electron.

Excerpt from Wikipedia:

In physics, the Lyman-alpha line is a spectral line of hydrogen, or more generally of one-electron ions, in the Lyman series, emitted when the electron falls from the n = 2 orbital to the n = 1 orbital, where n is the principal quantum number. In hydrogen, its wavelength of 1215.67 angstroms corresponding to frequency of 10^{15} hertz, places the Lyman-alpha line in the ultraviolet part of the electromagnetic spectrum, which is absorbed by air. Lyman-alpha astronomy must therefore ordinarily be carried out by satellite-borne instruments, except for extremely distant sources whose red shifts allow the hydrogen line to penetrate the atmosphere.

(Excerpt end)

Observation:

This wavelength is important because a quasar usually has this emission line in its spectrum. A shift of this emission line wavelength indicates the relative velocity of the atom.

10.4 Doppler Effect

The Doppler Effect is observed by the entire spectrum of the light source being shifted in proportion to the source's velocity in that direction.
The velocity of light is set by the medium. The velocity of light cannot be affected by the light source velocity. However, the source in motion affects the distribution of the radiated energy, not its velocity.

The timing of the Doppler Effect is crucial when one observes a spectrum shift in radiation from distant objects.

The Doppler Effect occurs only at the moment of radiation emission, when the motion of the object at that instant affects the spectrum.

There are 2 sources of electromagnetic radiation affected by the Doppler Effect: stars and atoms. Each initiates the propagation of the synchronized electric and magnetic fields. This propagation is an expanding sphere from the source. This sphere of energy continues until it is absorbed by an object in its path.

Stars emit a broad spectrum of thermal radiation.

Atoms emit a characteristic wave length based on the electron configuration.

The energy being lost in the atom is transferred to the corresponding wave lengths of electromagnetic radiation. Some atoms emit more than one wave length when dropping to their ground state.

These wave lengths can be observed and measured in a spectrum, and are called emission lines.

The instant of radiation emission, the motion of the source affects the wave length distribution around that sphere. Wave lengths in the direction of the source are changed by an amount proportional to the sources velocity relative to the velocity of light. The light source is generating a continuum of energy as a sphere. Wave lengths in one side of the sphere will be reduced, or toward the blue end, in the direction of the source. Wave lengths in the other side of the sphere will be increased, or toward the red end, in the direction opposite of the source. There is perfect symmetry with the change in wave length on one side exactly matched by the change on the opposite side. The sphere is a continuum of energy, being carried in wave lengths. There is definitely no quantized behavior present.

The motion of the light source does not change the amount of energy being radiated, only its distribution around the sphere of its propagation. Energy is always conserved.

The Doppler Effect also occurs only at the moment of radiation absorption, when the motion of the object at that instant also affects the spectrum. When energy is absorbed by an object than that energy is missing from the radiation. The energy is carried in wave lengths so those wave lengths carrying the energy which was transferred to the object are missing in the spectrum. These missing wave lengths are called absorption lines.

Absorption lines arise from objects in the line of sight, between the light source which emits the intact energy or spectrum.

The absorption line behavior is affected by the velocity of the atom. A moving atom carries kinetic energy and that energy participates in the transfer of energy from the radiation to the atom. As with an emission line, the velocity of the atom relative to the velocity of light determines the energy involved in the exchange.

An atom is essentially a tiny sphere. An atom in the path of electromagnetic radiation can absorb energy from that continuum of energy. The atom's motion relative to the radiation is important. The motion at that point in the sphere will have a proportion relative to the velocity of light and relative to the direction of the incoming light.

When the atom is moving toward the light source the kinetic energy of the atom is a participant and it reduces the energy the atom requires and absorbs from the radiation. This decrease in energy is a higher wave length.

Energy is always conserved during this exchange.

When the atom is moving away from the light source the kinetic energy of the atom is a participant and it increases the energy the atom requires and absorbs from the radiation. This increase in energy is a lower wave length.
The energy being absorbed is noted as an absorption line wave length.

This is the simple calculation of z.

The velocity, called v here, of the source is compared to the velocity of light by dividing that value by the velocity of light, called the constant c.

The value of v has a sign. Doppler Effect is in the observer's line of sight. When the object is moving away from the observer, v is + or positive, and when moving toward the observer, v is – or negative.

The result is called z by convention.

The simple equation is $z=v/c$, making sure the units are the same (usually km/s).

The shift in a spectrum due to the motion of the light source is a simple equation,
where EWL is the emission wavelength,

NWL is the new wavelength, so:
NWL = EWL + (EWL multiplied by z)

where the z is the factor for the change in the new wavelength from that originally emitted; z is positive for a red shift or negative for a blue shift.

There is no quantized behavior in any of the equation's factors or in the result.

10.4.1 Galaxy Red Shift

The spectrum of galaxies beyond our Local Group exhibit a unique behavior. In 1936, Edwin Hubble noticed this and put our Local Group on an island separate from the Hubble Flow.

These galaxies have an absorption line which shifts toward the red, and this shift is roughly proportional to the galaxy's distance from the observer, who is always on or near the Earth. This single line was attributed to hydrogen.

The explanation for the line shift is hydrogen atoms in the intergalactic medium (IGM), the space beyond our Local Group of galaxies. These atoms absorb its wave length, but then drop to their ground state and re-emit the wave length again.

This process of absorption and re-emission results in this observed increasing red shift based on the density of hydrogen atoms in the IGM. This red shift is caused by the IGM in the line of sight, not the light source.

By mistake, this hydrogen absorption line red shift was considered the result of a velocity causing a Doppler effect. This is only a line of sight behavior and indicates nothing about the distant galaxy's actual velocity. This mistake caused many others, including the universe expansion, dark energy, and the Big Bang.

Improving imaging technology enables a spectrum to be captured from galaxies which had been too dim by their distance.
Because this galaxy red shift is driven by the distance through the IGM, their red shift also increases by this measurement. Essentially, the only limit on a galaxy red shift is the technology to measure the most distant ones. Treating this z as a velocity of the galaxy is ridiculous. Scientists eventually tried to explain how galaxies could possibly have a velocity exceeding 8x the velocity of light. Their conclusion was the red shift must be coming from the IGM.

10.4.2 Quasar Red Shift

A quasar is a distant object which looks like a star but it has a strong source of synchrotron radiation, extending from radio to X-ray, All share red-shifted emission lines from a variety of non-hydrogen elements where the mix can vary by quasar. All share the same red shifted hydrogen Lyman-alpha emission line.

These quasars can have this line with a red shift indicating the atom is moving at many multiples of the speed of light, like 7x. A proton when capturing an electron emits this wave length. The wave length is shifted by the proton's velocity at the instant of that capture. This red shift comes from the atom in the line of sight, and indicates nothing about the distant quasar's actual velocity. This mistake compounds the galaxy red shift mistake, so both objects having a different mechanism make the false dark energy difficult to explain both false velocities.

Also, a quasar's hydrogen red shift of $z > 1$ indicates a proton's velocity is exceeding that of light. Einstein developed the theory of relativity assuming mass cannot travel faster than c. His assumption was shown to be a mistake by many quasars. Relativity has too many mistakes.

10.5 Synchrotron Radiation

Excerpt from Wikipedia:

Synchrotron radiation, electromagnetic energy emitted by charged particles (e.g., electrons and ions) that are moving at speeds close to that of light when their paths are altered, as by a magnetic field. It is so called because particles moving at such speeds in a variety of particle accelerator that is known as a synchrotron produce electromagnetic radiation of this sort.

Many kinds of astronomical objects have been found to emit synchrotron radiation as well. High-energy electrons spiraling through the lines of force of the magnetic field around the planet Jupiter, for example, give off synchrotron radiation at radio wavelengths. Synchrotron radiation at such wavelengths and at those of visible and ultraviolet light is generated by electrons moving in the magnetic field associated with the supernova remnant known as the Crab Nebula. Radio emissions of the synchrotron variety also have been detected from other supernova remnants in the Milky Way Galaxy and from extragalactic objects called quasars.

(Excerpt end)

Observation:

There are many X-ray point sources in the universe including one at the core of most spiral galaxies. These sources were described in detail in the author's book Cosmology Transition.

As somewhat described in the excerpt above, all those X-ray sources have an electrical current whose path is bent by a magnetic field resulting in this broad spectrum of wave lengths spanning from X-ray to infrared.

Quasars are typically dimmed in the optical wave lengths by their surrounding clouds of gas and dust.

10.6 Thermal Radiation

Excerpt from Wikipedia:

Thermal radiation is electromagnetic radiation generated by the thermal motion of particles in matter. All matter with a temperature greater than absolute zero emits thermal radiation.

If a radiation object meets the physical characteristics of a black body in thermodynamic equilibrium, the radiation is called blackbody radiation. Planck's law describes the spectrum of blackbody radiation, which depends solely on the object's temperature. Wien's displacement law determines the most likely frequency of the emitted radiation, and the Stefan–Boltzmann law gives the radiant intensity for the wave length.

(Excerpt end)

Thermal radiation is also one of the fundamental mechanisms of heat transfer. Conduction between adjacent solid objects is another.

Its spectrum is characterized by a wave length distribution, with the wave length having the highest intensity related to the object's temperature.

The wave length distribution affects whether it is visible. A cool temperature won't be. When warmer the increasing infrared intensity can be felt as heat or warmth but not seen. A rising temperature will become visible as red. When even hotter the mix of color wave lengths can result in "white hot." Our Sun is hot enough to generate the ultraviolet frequency which is not visible but can affect the eyes and skin.

Our white Sun can appear yellow when overhead due to the wave length distribution after the light passes through our atmosphere. The atmosphere can also cause a color change between sun rise and sun set, and it causes the sky to be blue.

Here is the thermal radiation spectrum from our Sun (from Wikipedia)

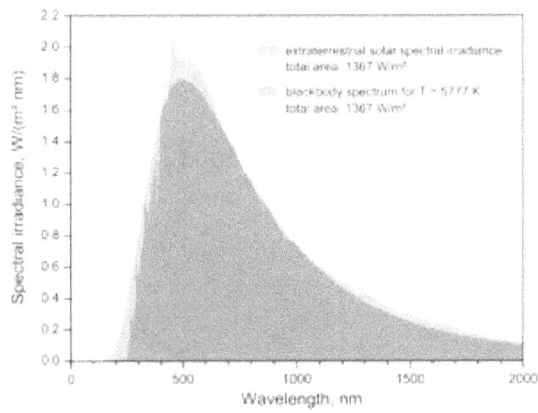

Important note about wave lengths:

Thermal radiation typically spans a continuum of energy from ultraviolet to infrared to wave lengths covering most temperatures.

Infrared is always present but shorter wave lengths arise with a high enough surface temperature. Our Sun's thermal radiation, seen as light, is in this wave length range of UV to infrared.

Most emission lines from atoms range from visible to ultraviolet wave lengths. As a general rule, any wave lengths measured outside of this range, like radio at the low end, and X-ray or gamma ray at the high end, were emitted by a source of synchrotron radiation.

A fictitious black hole violates this general rule because the impossible hot accretion disk is claimed to emit X-rays but that energy requires an impossible temperature.

Thermal radiation requires a surface.

The temperature of a gas is measured by the kinetic energy of its atoms or molecules. A gas cannot emit thermal radiation. When its atoms and molecules become ionized, then as each ion captures an electron, they emit their characteristic wave length of electromagnetic radiation. This is the non-thermal mechanism for the color of a neon light.

10.7 Red shift summary

The term "red shift" is used so loosely, most think of it as just a simple number having a consistent meaning, like a temperature.

A red shift is not that simple and anyone using the term so loosely is showing they consider it as just a simple number.

It is crucial to recognize there are 4 different red shifts. Each is a measurement of a distinct behavior.

Galaxies are totally different entities than quasars. A galaxy has billions of stars while a quasar is a quasi-stellar object having no stars.

A metallic element is one which is not hydrogen or helium.

The 4 distinct red shifts:

1) galaxy – hydrogen

2) galaxy – metal

3) quasar – hydrogen

4) quasar – metal

(1) the hydrogen absorption line is driven by hydrogen in the inter-galactic medium. This line is not from the galaxy.

(2) the calcium ion absorption line is driven by calcium ions near the galactic corona, as in the case of M31 and others. Calcium is a metal. The metallic line is not from the galaxy.

(3) The quasar high red shift comes from the hydrogen Lyman-alpha emission line.

(4) The quasar low red shift comes from the metallic ion emission lines.

(1) can never be a galaxy velocity. However, when used in conjunction with Cepheids, this value enables a distance calculation, with Cepheids providing a distance metric for the hydrogen density within the IGM in the line of sight to its galaxy.

(2) there are galaxies with either a red or blue shift of the metallic ion absorption lines. M31 has a calcium line blue shifted. This can never be a galaxy velocity, nor can it be related to a galaxy distance. Only a Cepheid provides a distance metric.

LINER galaxies, which include Seyferts, exhibit several metallic elements when taking the spectrum of only the AGN. None of these metallic lines in a LINER galaxy spectrum are related to the galaxy motion.

(3) this hydrogen emission line is found in a "typical" quasar. This can never be a quasar velocity, nor can it be related to a quasar distance.

(4) these metallic lines are found in the quasars used by Halton Arp, in his book Seeing Red. This can never be a quasar velocity, nor can it be related to a quasar distance, nor can it be related to the age of matter. These ions just slow down in apparent incremental changes in their velocity.

The z value for (3) has exceeded 7, while the z value for (4) is < 1.

It is crucial to note that none of the 4 types of a red shift is an indicator of the object's real velocity.

When one accepts that simple fact about the false velocities, then there is no "Hubble Flow." That was the term Edwin Hubble used initially for the red shift trend, but later in 1936, he noticed (1) is observed with only galaxies beyond our Local Group.

Hubble recognized the "Hubble Flow" was not consistent. Dark energy arose from the wrong assumption that the false expansion is consistent.

There is also no expansion, no dark energy, and no big bang.

10.8 Hierarchy of Reactions to light

All of an atom's reaction to light involve the nucleus and the innermost ring of 2 electrons.
An atom in the path of electromagnetic radiation must do one of the following:
1)   Particle pair production,
2)   Ionization,
3)   Compton scattering,
4)   Photoelectric effect,
5)   Absorption line,
6)   Reflect or re-emit it,
7)   Transfer to vibration in a molecular bond,
8)   Transfer to kinetic energy of the atom or molecule.

If (1) can be done, then the action is performed.
If not, the possible sequence of actions continues.

If (2) can be done, then the action is performed.
If not, the sequence continues.
If (3) can be done, then the action is performed.
If not, the sequence continues.
If (4) can be done, then the action is performed.
If (5) can be done, then the action is performed.
If (6) can be done, then the action is performed.
If (7) can be done, then the action is performed.
Action (7 or 8) must be done, if no other.

Energy must be transferred, as in (1,2,3.4,5,6,7), or transformed, as in (8).
Actions (1,2,5) absorb only some of the incoming energy but the total energy is conserved through the partial transfer.

Action (1) has two requirements. First, the atom's inner shell must have its 2 electrons because 2 will be ejected.

Descriptions of this action do not list the elements tested for this action.

The second requirement is the atom's state of matter. The description of its observations mention a "cloud chamber" implying this action has been observed only with unbound atoms in a gas.

Because the descriptions of this action lack all the necessary details, it is impossible to thoroughly explain its requirements. The mechanism can be explained. That explanation is in section 6.

Action (2) can occur in any state. If the radiation in its continuum has the energy for the outer shell to eject an electron, then in that instant, the energy is transferred to the ejected electron's kinetic energy.

Electromagnetic radiation continues its propagation until either absorbed or transferred.

Action (3) is described separately below. This is a behavior on a surface.

Action (4) is described separately below.

Action (5) occurs in an atom or molecule in a gas.

The atom can be neutral or ionized, but must have at least one electron. Doppler effect was explained earlier in this section 10.

Actions (3,4,6,7) can occur in a liquid or solid surface.

Action (8) is an instantaneous transfer of energy from the wave length to the particle's kinetic energy.

Thermal energy in a gas is held in the kinetic energy of its particles. Pressure and volume also affect the temperature of a gas.

10.9 Compton Scattering

Compton scattering is an atomic behavior involving the absorption of energy at a level between that required for particle pair production and the photoelectric effect.

Excerpt from Wikipedia:

Compton scattering, discovered by Arthur Holly Compton, is the scattering of a photon by a charged particle, usually an electron. If it results in a decrease in energy (increase in wavelength) of the photon (which may be an X-ray or gamma ray photon), it is called the Compton effect. Part of the energy of the photon is transferred to the recoiling electron. Inverse Compton scattering occurs when a charged particle transfers part of its energy to a photon. Compton found that some X-rays experienced no wavelength shift despite being scattered through large angles; in each of these cases the photon failed to eject an electron. Thus the magnitude of the shift is related not to the Compton wavelength of the electron, but to the Compton wavelength of the entire atom, which can be upwards of 10000 times smaller. This is known as "coherent" scattering off the entire atom since the atom remains intact, gaining no internal excitation.
In Compton's original experiments the wavelength shift given above was the directly-measurable observable. In modern experiments it is conventional to measure the energies, not the wavelengths, of the scattered photons.

(Excerpt end)

Observation:

This behavior is assumed to be an interaction between a photon and an electron, and claimed to be confirmation of light as a particle, the photon, and not a wave length.

This is an outright contradiction to Compton's conclusion the emitted wave length is determined by the "entire atom."

There is no photon interacting with a free charged particle. An atom's outer shell electrons are absorbing the energy required to change its energy to one which is exactly acceptable as defined by the atom and its electrons.

This behavior is the intermediate result between the other energy levels for the atom. Photoelectric effect results in an electron ejection, with the excess energy transferred to the departing electron.

The explanation for Compton scattering requires the atom absorb the necessary energy, for this action, among its electrons thereby increasing its energy level held among them. That energy must be released when the atom returns to its ground state.

Unlike the photoelectric effect having 1 action, at this higher energy level of Compton scattering, the atom performs 2 electron actions.

1) 1 electron is ejected.

2) A second electron moved to ground state resulting in the radiation for that change in its energy. This charge moved a very short distance resulting in a very short emission line.

The description states this wave length change is not quantized.

Excerpt from above:

"The wavelength shift is at least zero and at most twice the Compton wavelength of the electron."

The Wikipedia image and description of Compton's experiment in 1923 shows a "graphite target" suggesting a target's surface having a lattice of carbon Carbon has 6 electrons. There is no mention of a list of the other elements having this behavior measured with consistent results.

By comparison, the higher energy particle pair production (PPairP) also affects 2 electrons in the atom.
The difference between them is PPairP ejects the second electron as a positron, while CS gets a wave length, increased than that absorbed by the atom, from the non-ejected electron.

Compton scattering (CS) is a wave length not a photon behavior.

10.10 Particle Pair Production

Excerpt from Wikipedia for particle pair production:

Pair production often refers specifically to a photon creating an electron–positron pair near a nucleus. For pair production to occur, the incoming energy of the photon must be above a threshold of at least the total rest mass energy of the two particles, and the situation must conserve both energy and momentum. However, all other conserved quantum numbers (angular momentum, electric charge, lepton number) of the produced particles must sum to zero – thus the created particles shall have opposite values of each other. For instance, if one particle has electric charge of +1 the other must have electric charge of −1, or if one particle has strangeness of +1 then another one must have strangeness of −1.

The probability of pair production in photon–matter interactions increases with photon energy and also increases approximately as the square of atomic number of the nearby atom.

(Excerpts end)

Observation:

Particle pair production probability increases with more protons, so more energy can be absorbed by the heavier nucleus.

An atom will absorb a specific wavelength when its nucleus and inner pair of electrons to a new energy state by that amount being absorbed. This is a quantized behavior of an atom, where a longer wavelength, having less energy than required, will not be absorbed by the atom.

Similarly, when an electron moves to a lower orbital, or to a lower energy state, an emission line of a particular wave length is emitted. This wave length is sometimes related to the distance between orbitals and it contains the energy being released from the electron's change in its energy.

The photoelectric effect has an extra result with the absorption line, by ejecting an electron.
When the atom absorbs enough energy for an electron to leave the atom rather than just changing orbitals, then the electron departs having the kinetic energy with the excess over the minimum required to leave.

The pair production event description is awkward in the excerpt, with "[creating a] pair near a nucleus" when describing the event is actually changing an electron pair in orbit around the nucleus. Of course, that inner orbit is "near."

Therefore, this is the proposed mechanism:

The gamma wave length energy is much greater than the ultraviolet wave length energy for a single electron ejection.

The substantial additional energy being absorbed by the atom from the energy in one short wave length causes another particle ejection, except the second electron flips its charge's negative polarity to positive becoming a positron.

Currently, particle pair production is claimed to create matter from energy.

With this alternate mechanism, there is no matter created during the event. Instead, an electron changed to a positron. The event caused no change in mass in any particles. Also, then there is no known mechanism to convert energy into matter, where matter is usually protons and electrons, the two components of every atom.

This alternate mechanism solves the antimatter problem. Physicists cannot explain the lack of antimatter in the universe. They assume with no justification, matter and antimatter should have been created in similar quantities.

Antimatter is actually created by only by high velocity particles, from either particle accelerators like the LHC, or from cosmic rays. The one exception is the positron, or anti-electron, which can occur during an atom's radioactive decay.

The antiproton is created when a proton flips its charge polarity, just as an electron can do, as described above. Currently, an antiproton can be created only in very high energy, high velocity particle collisions. This was done in 1955 but the specific particles or nuclei being used in the collision were not identified. Sometimes the deuteron, which is proton + neutron, is mentioned for particle colliders because there are several convenient sources for hard water, but never explicitly about an antiproton. Both participants in a collision must have protons, because electrons have so little mass for this. The simple explanation is protons are just flipping their charge polarity, to become an antiproton. This is more believable than an antiproton somehow appearing.

The presence of protons when an antiproton appears is not a coincidence. That is the only mechanism for creating an antiproton.

These mechanisms for anti-particles are infrequent in the universe so, of course, the anti-particles are rare.

10.11 Photoelectric Effect

The photoelectric effect apparently resulted in the concept of a photon particle.

Excerpt from Wikipedia:

The photoelectric effect is the emission of electrons when electromagnetic radiation, such as light, hits a material. Electrons emitted in this manner are called photoelectrons.

The experimental results instead show that electrons are dislodged only when the light exceeds a certain frequency—regardless of the light's intensity or duration of exposure. Because a low-frequency beam at a high intensity could not build up the energy required to produce photoelectrons like it would have if light's energy was coming from a continuous wave, Albert Einstein proposed that a beam of light is not a wave propagating through space, but a collection of discrete wave packets, known as photons.

In 1905, Einstein proposed a theory of the photoelectric effect using a concept first put forward by Max Planck that light consists of tiny packets of energy known as photons or light quanta. Each packet carries hv energy that is proportional to the frequency v of the corresponding electromagnetic wave. The proportionality constant h has become known as the Planck constant.

The maximum kinetic energy K_{max} of the electrons that were delivered this much energy before being removed from their atomic binding is

$$K_{max} = hv - W$$

where W is the minimum energy required to remove an electron from the surface of the material.
Einstein's formula, however simple, explained all the phenomenology of the photoelectric effect, and had far-reaching consequences in the development of quantum mechanics.

(Excerpt end)

Observation:

Einstein described the phenomena but he did not justify a photon particle.

The energy requirement for absorption is defined by the atom. To properly test this effect, a continuum of energy must be provided The atom will absorb only the amount of energy required to complete its action, whether only enough energy to eject an electron, or accepting an excess which is passed to the electron as its kinetic energy. Energy is always conserved in the instant.

The quantized behavior is in the atom, not in the light.

This scenario is like a baby who accepts only a mouthful of milk from the bottle. The amount in a mouthful is defined by the baby, not by the milk or by the bottle.

Light is a continuous stream of energy not a collection of discrete wave packets.

Visible Light is a continuum of frequencies, essentially from violet to red. There are no discrete increments anywhere in this continuum of energy.

Our eyes see the combination of certain frequencies as white. Human eyes are not sensitive to only certain discrete packets.

There are no photons. Quantum mechanics just calls a wave length a photon.
However, wave lengths have no defined increment but span a continuum of values in whatever units are used, like Angstroms. The units selected for a measurement cannot define a behavior.

10.12 Reflection

After light is emitted, it continues its propagation until absorbed.

Some surfaces, having the lattice structure of condensed matter can absorb and re-emit the incoming energy.

This action is observed with the surface of water or glass.

A mirror has the reflective surface behind the transparent glass.

The transparent glass has a diffraction index so the velocity of light's propagation slows through the glass without being absorbed.

After passing through the glass, the propagation is affected by the current medium. When the back surface is not transparent, it must absorb or re-emit the light.

10.13 Molecular Vibration

When the energy in light is absorbed it can be transformed to thermal energy in condensed matter (liquid or solid but not gas).

Excerpt from Wikipedia:

A molecular vibration is a periodic motion of the atoms of a molecule relative to each other, such that the center of mass of the molecule remains unchanged.
The typical vibrational frequencies, range from less than 1013 Hz to approximately 1014 Hz, corresponding to wavenumbers of approximately 300 to 3000 cm−1.
In general, a non-linear molecule with N atoms has 3N − 6 normal modes of vibration, but a linear molecule has 3N − 5 modes, because rotation about the molecular axis cannot be observed. A diatomic molecule has one normal mode of vibration, since it can only stretch or compress the single bond. Vibrations of polyatomic molecules are described in terms of normal modes, which are independent of each other, but each normal mode involves simultaneous vibrations of different parts of the molecule.
A molecular vibration is excited when the molecule absorbs energy, ΔE, corresponding to the vibration's frequency, v, according to the relation $\Delta E = hv$, where h is Planck's constant. A fundamental vibration is evoked when one such quantum of energy is absorbed by the molecule in its ground state. When multiple quanta are absorbed, the first and possibly higher overtones are excited.

(Excerpt end)

Observation:

Conservation of energy is always maintained. When light encounters an atom, the energy must be absorbed, re-emitted, transferred or transformed.
Cosmological red shift has been proposed as an explanation celestial red shifts, caused by the expansion of the supposed fabric of space.. This is wrong because a spectrum change during the propagation of light violates thermodynamics. Every valid red or blue shift occurs at the moment of either emission or absorption. A spectrum change can occur at no other time.

11 Final Conclusion

Proposing a practical atomic model is the main purpose for this book. The word practical implies unnecessary concepts are discarded.

The probabilities and uncertainties offered in the Standard Model using quantum mechanics are not appropriate for the observed atomic behaviors involving electrons.

An electron is a fundamental particle having a measured mass, size, and negative charge. Electrons are real particles, having. They are subject to Coulomb's force between charged particles. In an atom, there are electrons in the nucleus, helping maintain stability with protons in contact or proximity.

The Standard Model is quite flawed when treating an electron as not a particle but as a wave function driven by probabilities.

Perhaps this perception is the reason the atomic mass defect remains without a suitable explanation. Atomic mass defect should be at a higher priority than assigned for the Standard Model. The defect arises from the real particle interactions in the nucleus because it involves the physical compression of a proton during fusion.

Electrons are in something of a cloud around the nucleus, in a defined sequence of circular rings, affected by both the positive charged nucleus and the other negative charged electrons. There is a balance among these individual electrical forces within an atom.

The Bohr model with its circular orbits loosely based on the solar system (having elliptical orbits) was a better description of an atom than the Standard Model just because it had electrons as real particles in real circular orbits. It is not clear whether it correctly explained the valence behavior in its 1913 version of the Bohr model.

The development of the spdf naming convention of the circular shells lead to using predictable circular rings having a capacity of 2, 6, 10, or 14 electrons. This model's evolution was nearing a model covering all the essentials. Unfortunately, the acceptance of relativity by about 1920, resulted in a diversion. Relativity enabled abstractions to be acceptable. Gravity is a real force and the abstraction of space-time is not a legitimate replacement. Whether by coincidence or not, around this time an electron changed from a particle to a wave-particle duality and later a wave function.

The current atomic model has the p, d, f orbitals in non-circular orbits. However, a circular orbit maintains a uniform force between the electron in orbit and the positive charges in the nucleus, which implies stability. Orbits having multiple lobes result in a varying force on the electron, which makes explaining its stability awkward and explaining the mechanism driving this varying force difficult.

This book demonstrates the electrons are in rings having a defined order and quantity for each element. Chemistry requires accurate data and predictable behaviors in each atom.

The measurement of ionization energies is as expected and without a stated margin of error which is required to account for any probability of electrons not being precisely at their predicted location in their circular ring around the nucleus.

The research in this book confirms the importance of an electron as a real particle, not a wave driven by probabilities.

There is no evidence supporting a theory that electrons are moving in any other path around a nucleus than a circle.

Protons make up most of the mass in an atom. Electrons are required in the nucleus to maintain equilibrium among the isostatic forces holding the protons together.

The combination of protons and electrons achieving the nucleus of all 252 nucleon counts with the longest longevity was identified.

Electrons also enable molecular bonds by being shared by 2 nuclei because the positive nuclei are attracted to the negative electrons in the middle while the 2 nuclei, at a further distance between them, repel each other.

The balance of these electrostatic forces between charges maintains stability of the chemical bond.

If the number of electrons is unable to maintain stability, the action taken in the nucleus toward stability is either: a) ejecting excess electrons, by a process called beta decay, or b) by capturing 1 or 2 electrons from the inner ring to supplement the electrons who are too few in number, or c) an alpha particle ejection when a combination of 4 protons and 2 electrons is ejected from the nucleus at high velocity due to the force of repulsion between protons, a force which the presence of attached electrons can temporarily suppress; this ejection is a big change in the unstable nucleus.

Electrons also provide the function of keeping lattice structures stable where atomic nuclei are at the nodes of the lattice and free electrons circulate to maintain the lattice.

This practical atomic model avoids the unnecessary naming confusion of the shells in the Standard Model based on quantum-based spdf orbitals having a varying number of lobes.

There are 20 elements having 1 or 2 electrons moved between adjacent circular rings in their configuration. The position change between circular rings is easily explained. Explaining such a transfer between quantum non-circular orbitals which are driven by a probability and uncertainty is awkward.

The practical atomic Model has the evidence for claiming electrons are in predictable circular rings and for describing proton and electron combinations to create the observed nuclei.

The topology of the nucleus affects its stability where an odd or even count of protons affects the number of electrons required for the atom's stability or longevity before decay.

The author created several spreadsheets to demonstrate these predictable behaviors in a nucleus and in the electron configuration around it. With data for each of the 118 elements and many of their isotopes, there are too much data to compress into a book. However, links are provided to the reader if the original data are desired for viewing outside of this book.

For quantum mechanics to persist, it must provide evidence for any claim. By just stating something is uncertain cannot be accepted unless somehow such a vague theory has evidence to confirm it has some value.

Proper science requires testing and evidence. There is no evidence for a prediction when a predicted outcome is claimed to be uncertain. Even a probability requires justification.

At this point, the proposed pdf quantum orbitals having electrons in noncircular orbits around the nucleus have no evidence. That part of an atomic model must be changed to suit the evidence. If the theory has any value, then it must be revised to fix the facets which do not match the evidence.

Unverified theories are useful for future research but they are inappropriate for inclusion when describing our current understanding, which must be based on the currently known evidence.

By naming the rings according to both their order and capacity, an element's electron configuration list is immediately useful. That usefulness is part of a practical atomic model.

12 References

The references in the book are available as clickable links from a page in the author's web site.

1. Start web browser

2. Go to this site: www.cosmologyview.com

3. Make sure the browser is on the correct home page:

Cosmology Views

4. Scroll to near the middle.

5. Select: **Books by the author**

This page presents information for each book.

Locate the rows and columns for this book.

6. Locate: **Practical Atomic Model**

7. Below it, locate the date of this book's edition:

Feb. 10, 2021 References

8. Select: **References** link after the correct date.

9. This selection presents a web page.

The page will list the references in the book by page number, with a link to that reference.

Each link indicates whether it is to a pdf, a YouTube video, or a URL link to a web page. The user is aware of what the browser will do with the link. Some browsers do a download of a pdf before its display.

www.ingramcontent.com/pod-product-compliance
Lightning Source LLC
Chambersburg PA
CBHW070324220526
45467CB00001B/25